William Henry Drew

A Geometrical Treatise on Conic Sections

with numerous examples for the use of schools and students in the universities: with an appendix on harmonic ratio, poles and polars, and reciprocation. Seventh Edition

William Henry Drew

A Geometrical Treatise on Conic Sections
with numerous examples for the use of schools and students in the universities: with an appendix on harmonic ratio, poles and polars, and reciprocation. Seventh Edition

ISBN/EAN: 9783337323486

Printed in Europe, USA, Canada, Australia, Japan

Cover: Foto ©berggeist007 / pixelio.de

More available books at **www.hansebooks.com**

A GEOMETRICAL TREATISE
ON
CONIC SECTIONS.

WITH NUMEROUS EXAMPLES.

For the Use of Schools and Students in the Universities.

WITH

AN APPENDIX ON HARMONIC RATIO, POLES AND POLARS, AND RECIPROCATION.

BY THE

REV. W. H. DREW, M.A.,

ST. JOHN'S COLLEGE, CAMBRIDGE,
PROFESSOR OF MATHEMATICS IN KING'S COLLEGE, LONDON.

SEVENTH EDITION.

London:
MACMILLAN AND CO.
1883.

[The Right of Translation and Reproduction is Reserved.]

PREFACE TO THE FIFTH EDITION.

In this Edition an Appendix has been added, in which an endeavour has been made to present the subject of Harmonic Ratio, Poles and Polars, and Reciprocation, in a form adapted to the wants of Students who approach these ideas for the first time.

A further collection of Problems has also been given, taken from Examination Papers of recent dates, and the work has thus been brought completely up to the requirements of the present time.

W. H. DREW.

King's College, London,
June 26, 1875.

CONTENTS.

INTRODUCTION . . . .

CHAPTER I.

THE PARABOLA . . . .

    PROBLEMS ON THE PARABOLA . . . . .

CHAPTER II.

THE ELLIPSE . . . .

    PROBLEMS ON THE ELLIPSE . . . .

CHAPTER III.

THE HYPERBOLA . . . .

    PROBLEMS ON THE HYPERBOLA . . . .

CHAPTER IV.

THE SECTIONS OF THE CONE . . . . . .

    PROBLEMS ON THE SECTIONS OF THE CONE . .

ADDITIONAL PROBLEMS . . . .

SECOND SERIES . . . .

APPENDIX . . . .

CONIC SECTIONS.

INTRODUCTION.

1. DEF. The curve traced out by a point, which moves in such a manner that its distance from a given fixed point continually bears the same ratio to its distance from a given fixed line, is called a *Conic Section*.

The fixed point is called the *Focus*, and the fixed line the *Directrix*.

Thus if S be the focus, and KK' the directrix, and P a point from which PM is drawn at right angles to the directrix, the curve traced out by P will be a *Conic Section*, provided P move in such a manner that SP always bears the same ratio to PM.

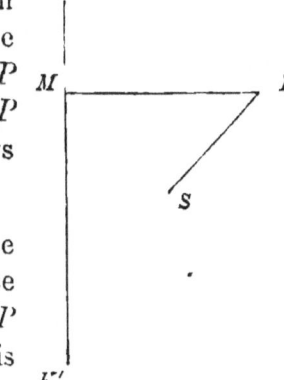

(1.) When the distance from the fixed point is equal to the distance from the fixed line, that is, when SP is equal to PM, the *Conic Section* is called a *Parabola*.

(2.) When the distance from the fixed point is less than the distance from the fixed line, that is, when the ratio which

SP bears to PM is less than unity, the *Conic Section* is called an *Ellipse*.

(3.) When the distance from the fixed point is greater than the distance from the fixed line, that is, when the ratio which SP bears to PM is greater than unity, the *Conic Section* is called an *Hyperbola*.

2. The reason of the term *Conic Sections* being applied to these curves is that, when a *Cone* is intersected by a plane surface, the boundary of the section so formed will, *in general*, be one or other of these curves.

I propose to investigate the properties of the *Conic Sections* from the definitions given above, and afterwards to show in what manner a *Cone* must be divided by a plane in order that the curve of intersection may be a *Parabola, Ellipse,* or *Hyperbola.*

CHAPTER I.

THE PARABOLA.

Prop. I.

3. **The focus and directrix of a parabola being given, to find any number of points on the curve.**

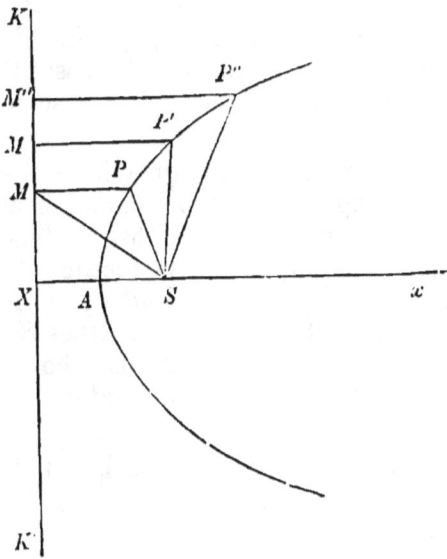

Let S be the focus, and KK' the directrix.

Draw XSx at right angles to the directrix, and bisect the line SX in A; then

$$\text{since } AS = AX,$$

∴ A is a point on the curve.

The point A is called the *Vertex*, and the line Ax, with respect to which the curve is evidently symmetrical, is called the *Axis*.

On the directrix take any point M; join SM; and draw MP at right angles to the directrix.

At the focus S make the angle MSP equal to the angle SMP; then
$$SP = PM,$$
$\therefore P$ is a point on the curve.

So by taking any number of points, M', M'', on the directrix, we may obtain as many points, P', P'', on the curve as we please, and the line which passes through A and all these points will be the parabola whose focus is S and directrix KK'.

Cor. 1. As M is taken further away from the point X, the line SM and the angles SMP, MSP, and, consequently, the lines SP and PM, continually increase. Hence, since XM and MP increase together, the curve recedes at the same time both from the axis and directrix; and since the angle SMP can never exceed a right angle, and the lines SP and MP will therefore always meet, it is evident that there is no limit to the distance to which the curve may extend on both sides of the axis.

Cor. 2. The parabola may be described practically in the following manner.

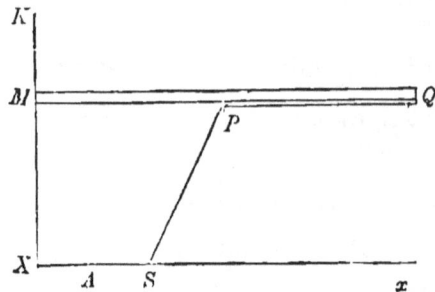

Let S be the focus and KX be the directrix; and let a rigid bar QM, having a string of the same length as itself fastened at one end Q, be made to slide parallel to the axis with the other end M on the directrix; then if the other end of the string be fastened at the focus, and the string be kept stretched by means of the point of a pencil at P, in contact with the bar, since SP will always be equal to PM, it is evident that the point P will trace out the parabola.

Prop. II.

4. The distance of any point inside the parabola from the focus is less than its distance from the directrix; and the distance of any point outside the parabola from the focus is greater than its distance from the directrix.

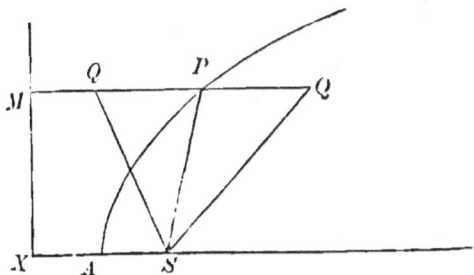

(1.) Let Q be a point inside the parabola.

Draw QM at right angles to the directrix, meeting the parabola in P; join SP; then

since $SP = PM$,

∴ SP and $PQ = QM$.

But SP and $PQ > SQ$,

∴ $QM > SQ$.

(2.) Let Q be a point outside the parabola.

Draw MQ at right angles to the directrix, and produce it to meet the parabola in P; join SP; then

since SQ and $QP > SP$,

and $SP = PM$,

∴ SQ and $QP > PM$,

∴ $SQ > QM$.

Cor. Conversely a point will be inside or outside the parabola according as its distance from the focus is less or greater than its distance from the directrix.

5. Def. The line PN (*see fig. Prop.* III.) drawn at right angles to the axis from the point P in the curve is called the *Ordinate* of the point P, and the line AN the *Abscissa*. The double ordinate BC drawn through the focus, and terminated both ways by the curve, is called the *Latus Rectum*.

Prop. III.

The Latus Rectum $BC = 4AS$.

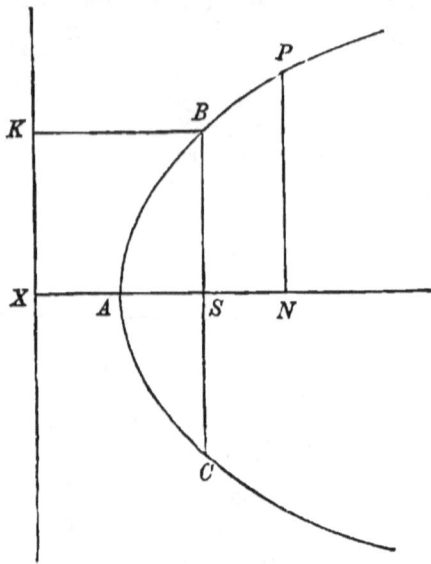

Draw BK at right angles to the directrix.

Then $SB = BK = SX = 2AS$,

$\therefore BC = 4AS$.

6. DEF. If a point P' be taken on the parabola (see *fig. Prop.* IV.) near to P, and PP' be joined, the line PP' produced, in the limiting position which it assumes when P' is made to approach indefinitely near to P, is called the *Tangent* to the parabola at the point P.

Prop. IV.

If the tangent to the parabola at any point P intersect the directrix in the point Z; then SZ will be at right angles to SP.

Let P' be a point on the parabola near to P.

CONIC SECTIONS.

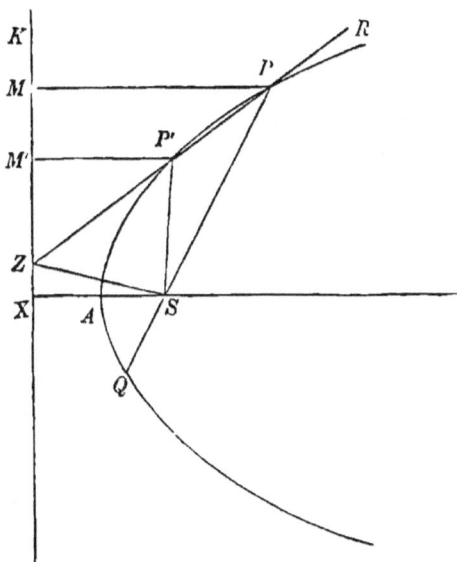

Draw the chord PP' and produce it to meet the directrix in Z; join SZ.

Draw PM, $P'M'$ at right angles to the directrix; join SP, SP'; and produce PS to meet the parabola in Q.

Then, since the triangles ZMP, $ZM'P'$ are similar,
$$\therefore ZP : ZP' :: MP : M'P',$$
$$:: SP : SP',$$
$\therefore SZ$ bisects the angle $P'SQ$. (*Euclid*, VI. *Prop.* A.)

Now when P' is indefinitely near to P, and PP' becomes the tangent at the point P, the angle PSP' becomes indefinitely small, while the angle QSP' approaches two right angles, and therefore the angle $P'SZ$, which is half of the angle $P'SQ$, becomes ultimately a right angle.

Hence, when PZ is the tangent,

the angle ZSP is a right angle,
or SZ is perpendicular to SP.

Cor. Conversely, if SZ be drawn at right angles to SP, meeting the directrix in Z, and PZ be joined, PZ will be a tangent at P.

Prop. V.

7. The tangent at any point P of a parabola bisects the angle between the focal distance SP, and the perpendicular PM on the directrix.

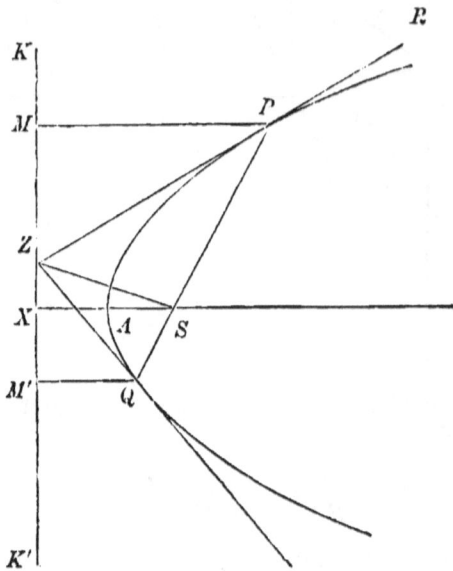

Let the tangent at P meet the directrix in the point Z; join SZ; then since the angle ZSP is a right angle, (*Prop.* IV.)

$$\therefore ZS^2 + SP^2 = PZ^2.$$
$$\text{Also } ZM^2 + MP^2 = PZ^2.$$
$$\therefore ZS^2 + SP^2 = ZM^2 + MP^2.$$
$$\text{But } SP = PM,$$
$$\therefore ZS = ZM.$$

Now in the triangles ZPS, ZPM,
$$\therefore ZP, PS = ZP, PM, \text{ each to each,}$$
$$\text{and } ZS = ZM,$$
$\therefore$ the angle $SPZ =$ the angle MPZ;
or PZ bisects the angle SPM.

COR. 1. If ZP be produced to R, then the angle $SPR =$ the angle MPR.

COR. 2. It is evident that the tangent at the vertex A is perpendicular to the axis.

PROP. VI.

8. The tangents at the extremities of a focal chord intersect at right angles in the directrix.

Let PSQ be a focal chord, and let the tangent at P meet the directrix in Z.

Join SZ; then

the angle ZSP is a right angle, (*Prop.* IV.)

and $\therefore$ also the angle ZSQ is a right angle,

$\therefore ZQ$ is the tangent at Q, (*Prop.* IV. *Cor.*)

or the tangents at the extremities of the focal chord PSQ intersect in the directrix.

Again, draw PM, QM' at right angles to the directrix; then

since $MP, PZ = SP, PZ$, each to each,

and the angle MPZ = the angle SPZ,

$\therefore$ the angle MZP = the angle SZP,

$\therefore$ the angle SZP is half of the angle SZM.

So the angle SZQ is half of the angle SZM',

$\therefore$ the angle PZQ is half of the two SZM and SZM'.

But the angles SZM and SZM' = two right angles,

$\therefore$ the angle PZQ is a right angle,

or the tangents at the extremities of a focal chord intersect *at right angles* in the directrix.

Prop. VII.

9. If the tangent at any point P of a parabola meet the axis produced in the point T, and PN be the ordinate of the point P, then $NT = 2AN$.

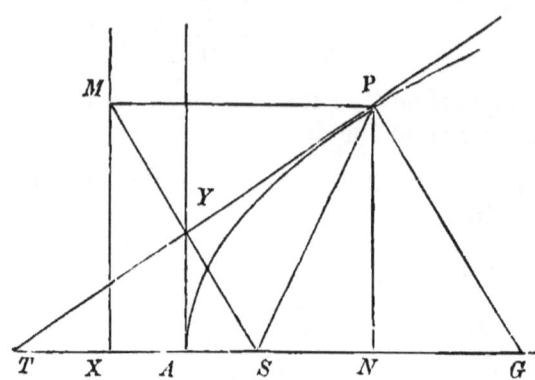

Join SP, and draw PM at right angles to the directrix; then

∴ the angle SPT = the angle MPT = the angle STP,
∴ $ST = SP$.
But $SP = PM = XN$,
∴ $ST = XN$.
But $AS = AX$,
∴ the remainder AT = the remainder AN,
∴ $NT = 2AN$.

DEF. The line NT is called the *Subtangent*.

10. DEF. The line PG, drawn at right angles to PT, is called the *Normal* at the point P, and NG the *Subnormal*.

Prop. VIII.

If the normal at the point P of a parabola meet the axis in the point G, then $NG = 2AS$.

Since the angle SPG = the complement of the angle SPT,
and the angle SGP = the complement of the angle STP,
and also the angle SPT = the angle STP, (*Prop.* VII.)
$\therefore$ the angle SPG = the angle SGP,
$\therefore SG = SP$.
But $SP = PM = XN$,
$\therefore SG = XN$.
Taking away the common part SN,
the remainder $NG = SX = 2AS$.

Prop. IX.

11. If PN be an ordinate to the parabola at the point P; then $PN^2 = 4AS \cdot AN$.

Since TPG is a right angle, and PN perpendicular to TG;
$\therefore PN$ is a mean proportional between TN and NG;
or $PN^2 = TN \cdot NG$. (*Euclid*, VI. 8 *Cor.*)
But $TN = 2AN$, (*Prop.* VII.)
and $NG = 2AS$, (*Prop.* VIII.)
$\therefore PN^2 = 4AS \cdot AN$.

Prop X.

12. If the tangent at any point P intersect the tangent at the vertex in Y, then SY will bisect PT at right angles, and will be a mean proportional between SA and SP.

Draw PN at right angles to the axis; then
since AY is parallel to PN,
$\therefore TY : YP :: TA : AN$.
But $AT = AN$, (*Prop.* VII.)
$\therefore TY = PY$;
and $\therefore SY, YP = SY, YT$, each to each,
and $SP = ST$, (*Prop.* VII.)
$\therefore$ the angle SYP = the angle SYT,
$\therefore SY$ is perpendicular to PT.

Again, since TYS is a right angle, and YA perpendicular to ST,

∴ SY is a mean proportional between ST and SA ;

or $SY^2 = ST \cdot SA$. (*Euclid*, VI. 8 *Cor.*)

But $ST = SP$, (*Prop.* VII.)

∴ $SY^2 = SP \cdot SA$.

Cor. If PM be drawn at right angles to the directrix, and MY be joined, then

since $SP, PY = MP, PY$, each to each,

and the angle $SPY =$ the angle MPY, (*Prop.* V.)

∴ the angle $SYP =$ the angle MYP,

∴ SY and YM are in the same straight line.

Prop. XI.

13. To draw a pair of tangents to a parabola from an external point.

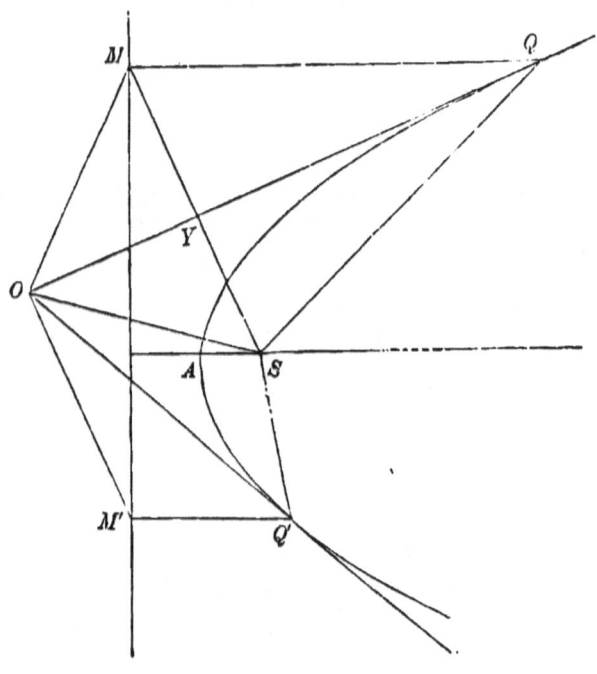

Let O be the given external point.

Join OS, and with centre O and radius OS describe a circle, cutting the directrix in M and M', which it will always do, on whichever side of the directrix O is situated, since O is nearer to the directrix than to the focus. (*Prop.* II.)

Draw MQ and $M'Q'$ parallel to the axis meeting the parabola in Q and Q'.

Join OQ, OQ'; these will be the tangents required.

Join SQ and SQ'; then

∴ OQ, $QS = OQ$, QM, each to each,

and $OS = OM$,

∴ the angle OQS = the angle OQM,

∴ OQ is the tangent at Q. (*Prop.* V.)

So OQ' is the tangent at Q'.

Prop. XII.

14. If from a point O a pair of tangents OQ and OQ' be drawn to a parabola, the triangles OSQ, OSQ' will be similar, and OS will be a mean proportional between SQ and SQ'.

Join SM, cutting OQ at right angles (*Prop.* X. *Cor.*) in the point Y; then

since the angle SQO = the angle MQO, (*Prop.* V.)

and the angle MQO = the angle SMM',

each of these angles being the complement of the angle QMY,

∴ the angle SQO = the angle SMM'.

But the angle SMM' at the circumference is half the angle SOM' at the centre, and is therefore equal to the angle SOQ'.

∴ the angle SQO = the angle SOQ'.

So the angle SOQ = the angle $SQ'O$,

∴ the remaining angle OSQ = the remaining angle OSQ'.

And therefore the triangle OSQ is similar to the triangle OSQ'

$$\therefore SQ : SO :: SO : SQ',$$
$$\therefore SQ . SQ' = SO^2.$$

or SO is a mean proportional between SQ and SQ'.

Prop. XIII.

15. If a pair of tangents OQ, OQ' be drawn to a parabola, and OV be drawn parallel to the axis meeting QQ' in V, then QQ' shall be bisected in V.

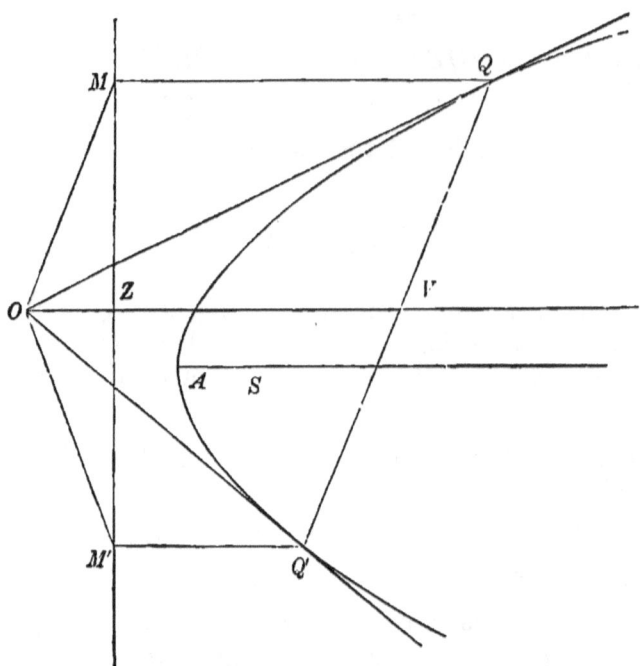

Draw QM, $Q'M'$ at right angles to the directrix. Join OM, OM'; and let OV meet MM' in Z.

Then, since $OM = OM'$, (*Prop.* XI.)
$\therefore$ the angle OMZ = the angle $OM'Z$,
and the angle OZM = the angle OZM',

CONIC SECTIONS. 15

and the side OZ is common to the triangles OZM, OZM',

$$\therefore MZ = M'Z.$$

And because the lines QM, ZV, $Q'M'$ are parallel,

$$\therefore QV : Q'V :: MZ : M'Z.$$

But $MZ = M'Z$,

$$\therefore QV = Q'V,$$

$\therefore QQ'$ is bisected in V.

Prop. XIV.

16. If from a point O a pair of tangents OQ, OQ', be drawn to a parabola, and OV be drawn parallel to the axis meeting the parabola in P, and QQ' in V, then the tangent at P will be parallel to QQ' and OV will be bisected in P.

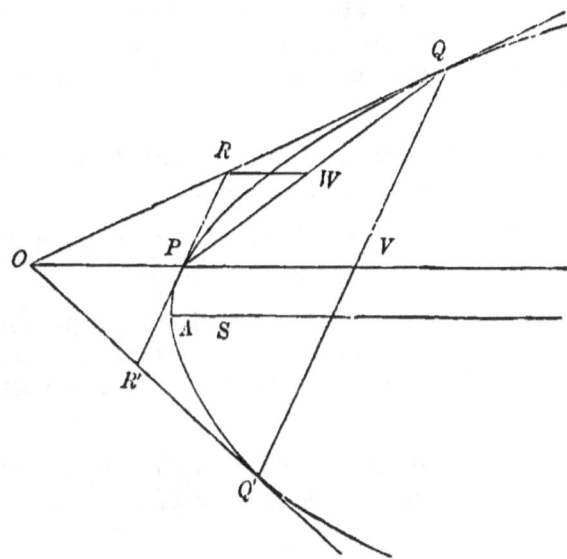

Draw the tangent RPR' meeting OQ, OQ' in R and R'.

Join PQ, and draw RW parallel to the axis, meeting PQ in W;

Then, by the last Proposition,
$$PW = WQ.$$
And because RW is parallel to OP,
$$\therefore OR : RQ :: PW : WQ.$$
But $PW = WQ$,
$$\therefore OR = RQ;$$
so $OR' = R'Q'$,
$$\therefore OR : RQ :: OR' : R'Q',$$
$$\therefore RR' \text{ is parallel to } QQ'.$$
Again, since PR is parallel to QV,
$$\therefore OP : PV :: OR : RQ.$$
But $OR = RQ$,
$$\therefore OP = PV.$$

Cor. From this it is manifest that if any number of parallel chords be drawn in a parabola, their middle points will all lie on the line parallel to the axis which passes through the point where the tangent drawn parallel to the chord meets the parabola.

Def. Any line PV, drawn from a point P in the parabola parallel to the axis, is called a *Diameter*.

The point P is called the *Vertex* of the diameter PV; and the tangent at P the *Tangent at the Vertex*.

The diameter consequently bisects all chords parallel to the tangent at the vertex, and the tangents at the extremities of any chord will intersect in the diameter corresponding to that chord.

Def. A line QV, drawn parallel to the tangent at P from a point Q in the curve, is called the *Ordinate* to the diameter PV.

Prop. XV.

17. If QV be an ordinate to the diameter PV, then $QV^2 = 4 . SP . PV$.

CONIC SECTIONS. 17

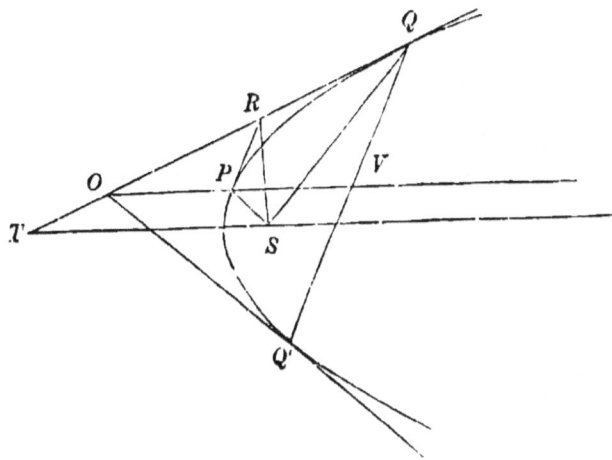

Produce QV to meet the parabola in Q'; and draw the tangents QO, $Q'O$, meeting VP produced in the point O. (*Prop*. XIV.)

Also let the tangent at P meet OQ in R, and join SP, SR, and SQ. Now since from the point R two tangents RP, RQ are drawn to the parabola, the triangle RPS is similar to the triangle RSQ, (*Prop*. XII.)

∴ the angle SRP = the angle SQR.

But the angle SQR = the angle STQ, (*Prop*. VII.)

= the angle POR,

∴ the angle SRP = the angle POR,

and the angle SPR = the angle OPR, (*Prop*. V. *Cor*. 1.)

∴ the remaining angle RSP = the remaining angle ORP,

∴ the triangle SPR is similar to the triangle POR.

∴ $SP : PR :: PR : PO$,

∴ $PR^2 = SP \cdot PO$,

= $SP \cdot PV$. (*Prop*. XIV.)

Again, since QV is parallel to PR,

∴ $QV : PR :: OV : OP$.

But $OV = 2OP$, (*Prop*. XIV.)

C

$$\therefore QV = 2PR,$$
$$\therefore QV^2 = 4PR^2,$$
$$= 4SP \cdot PV.$$

18. DEF. The double ordinate to the diameter PV, drawn parallel to the tangent at P, and passing through the focus, is called the *Parameter* of *the diameter PV*.

PROP. XVI.

The parameter of the diameter $PV = 4 \cdot SP$.

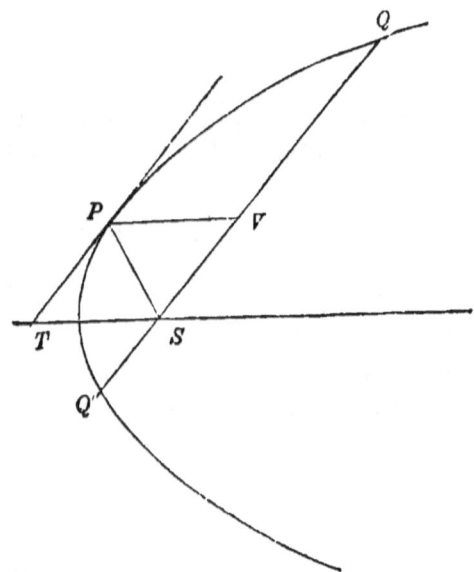

Draw QSQ' through the focus parallel to the tangent at P, and let the tangent at P meet the axis produced in T; then

$$QV^2 = 4\ SP \cdot PV. \quad (\textit{Prop. XV.})$$
But $PV = ST = SP$, (*Prop.* VII.)
$$\therefore QV^2 = 4SP^2;$$
or $QV = 2SP$,
$$\therefore QQ' = 4SP.$$

Prop. XVII.

19. If two chords of a parabola intersect one another, the rectangles contained by their segments are in the ratio of the parameters of the diameters which bisect the chords.

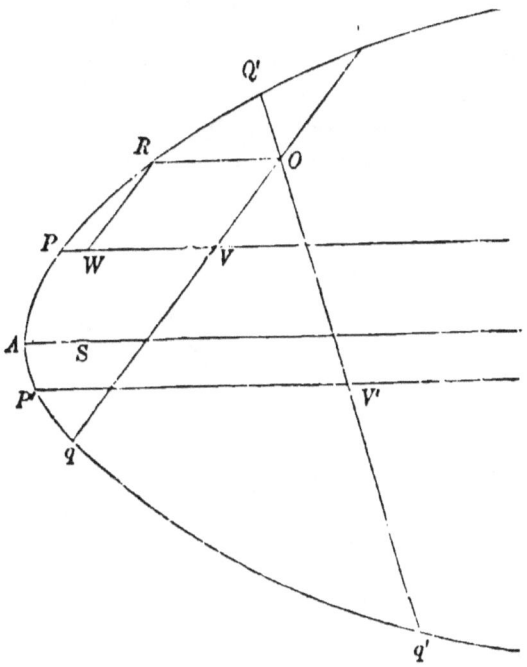

Let the chords Qq, $Q'q'$ intersect one another in the point O.

Bisect Qq, $Q'q'$ in V and V'; and draw the diameters PV, $P'V'$ parallel to the axis.

Also, through O draw OR parallel to PV; and through R draw RW parallel to QV.

Now, since Qq is divided equally in V and unequally in O,

$$\therefore QO \cdot Oq = QV^2 - OV^2, \quad (Euclid,\ \text{II. 5})$$
$$= QV^2 - RW^2,$$
$$= 4SP \cdot PV - 4SP \cdot PW, \quad (Prop.\ \text{XV.})$$
$$= 4Sp \cdot RO.$$

So $Q'O \cdot Oq' = 4SP' \cdot RO$.

Hence $QO \cdot Oq : Q'O \cdot Oq' :: 4SP : 4SP'$.

By *Euclid*, II. 6, the same may be proved to be true if the point O be without the parabola.

Prop. XVIII.

20. If from an external point O a pair of tangents OQ, OQ' be drawn to the parabola, and the chord QQ' be joined, the area of the figure bounded by QQ' and the curve is two-thirds of the triangle QOQ'.

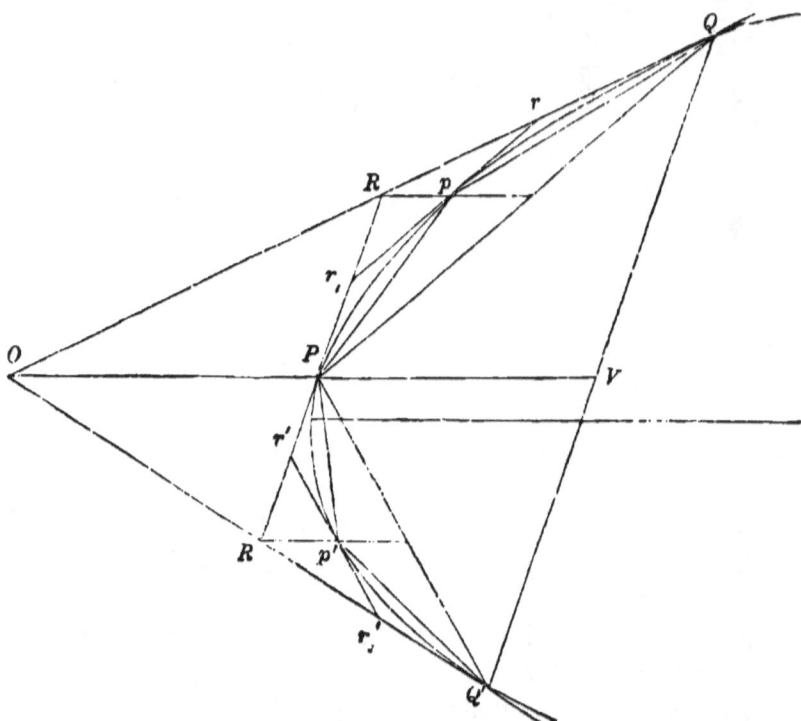

Draw the diameter OV meeting the curve in P; and let the tangent at P meet OQ, OQ' in R and R'.

Join QP, $Q'P$; then

since $OR = RQ$,

∴ the triangle $OPR = \frac{1}{2}$ the triangle OPQ,

$= \frac{1}{2}$ the triangle VPQ.

CONIC SECTIONS.

So the triangle $OPR = \frac{1}{2}$ the triangle VPQ',

∴ the triangle $ORR' = \frac{1}{2}$ the triangle PQQ'.

Again, if through R and R' we draw the diameters Rp, $R'p'$; and at the points p and p' draw the tangents $rpr_{,}, r'p'r'_{,}$ we can prove in the same manner as before that

the triangle $Rrr_{,} = \frac{1}{2}$ the triangle QpP,

and the triangle $R'r'r'_{,} = \frac{1}{2}$ the triangle $Q'p'P$.

Continuing in this manner to form new triangles by drawing diameters at the points r, $r_{,}$ and $r'r'_{,}$ and tangents at the points where these diameters meet the curve, we can prove that the exterior triangles formed by the tangents are the halves of the interior triangles formed by joining the points of contact with the extremities of the chords.

And the same will hold however the number of the triangles be increased.

Hence the sum of all the exterior triangles will be equal to half the sum of all the interior triangles.

Now when the number of the triangles is increased indefinitely, the sum of the exterior triangles will represent the exterior figure $OQPQ'$, and the sum of the interior triangles the area of the interior figure QPQ. Hence

the area of the figure $OQPQ' = \frac{1}{2}$ the area of the figure QPQ'

∴ area of the figure $OQPQ' = \frac{1}{3}$ the area of triangle QOQ',

∴ area of the figure $QPQ' = \frac{2}{3}$ the area of triangle QOQ'.

21. DEF. If with a point O on the normal at P as centre and OP as radius, a circle be described touching the parabola at P and cutting it in Q; then when the point Q is made to approach indefinitely near to P, the circle is called the *Circle of Curvature* at the point P. (See fig. Prop. XIX.)

Prop. XIX.

The chord of the circle of curvature, at a point P of a parabola, drawn parallel to the axis $= 4SP$.

Let PT be the tangent, and PG the normal at the point P.

With centre O and radius OP describe a circle cutting the parabola in the point Q.

Draw RQX parallel to the axis meeting the circle in X and the tangent at P in R.

Also draw QV parallel to PR, and PW parallel to the axis; then

since RP touches the circle at P,

$\therefore RQ \cdot RX = PR^2$. (*Euclid*, III. 36.)

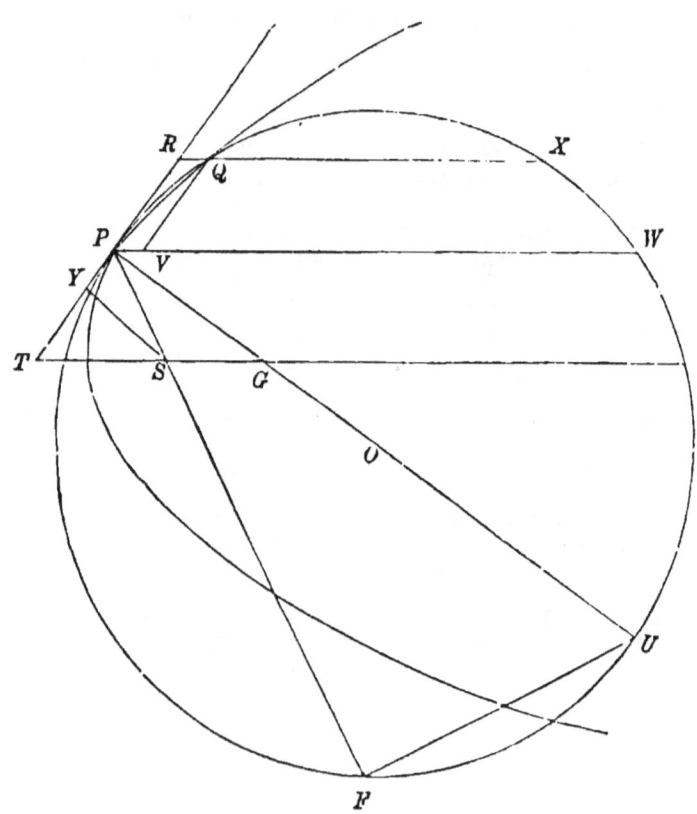

But $PR^2 = QV^2 = 4SP \cdot PV$, (*Prop.* XV.)
$\therefore RQ \cdot RX = 4SP \cdot PV$.
But $RQ = PV$.
$\therefore RX = 4SP$.

Now when the circle becomes the circle of curvature at P, the points R and Q move up to and coincide with P, and the lines RX and PW become equal.

Hence the chord of the circle of curvature parallel to the axis $= 4SP$.

Cor. 1. If PU be the diameter of the circle of curvature, and PF the chord through the focus; then

since the angle $FPU =$ the angle WPU, (*Prop.* VIII.)
$\therefore PF = PW = 4SP$.

Cor. 2. If SY be drawn at right angles to PT; then
the triangle PFU is similar to SYP,
$\therefore PU : PF :: SP : SY$,
or $PU : 4SP :: SP : SY$.

Prop. XX.

If QVQ' be any ordinate to the diameter PV, the circle described through the three points P, Q, Q' will intersect the parabola in a fourth point, which depends only upon the position of P.

Draw the ordinate PN, and produce it to meet the parabola in P'; then,

since the subtangent $= 2 \cdot AN$. (*Prop.* VII.)

the tangents at P and P' will meet the axis in the same point T.

Draw PR parallel to TP', meeting the parabola in R, and QQ' in O; then

$PO \cdot OR : QO \cdot OQ' :: SP' : SP$. (*Prop.* XVII.)

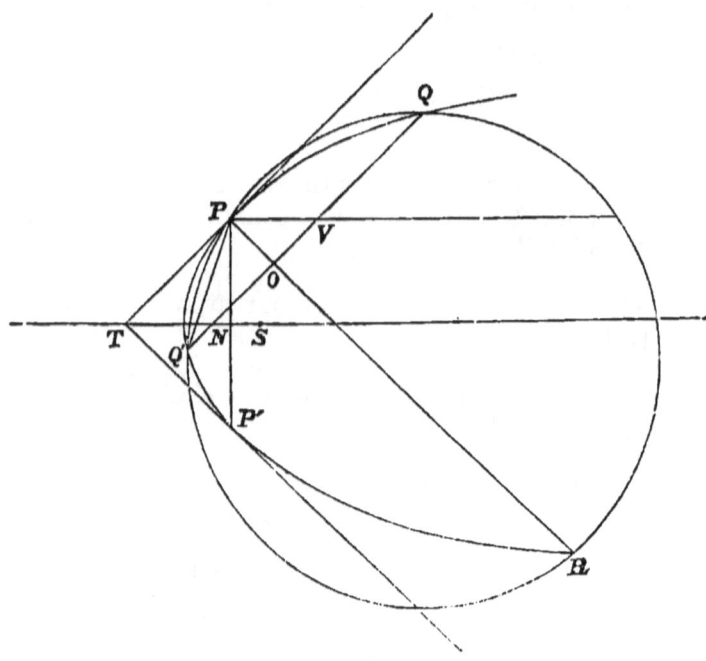

But $SP = ST = SP'$, (*Prop.* VII.)

$$\therefore PO \cdot OR = QO \cdot OQ'.$$

Hence by the converse of *Euclid* III. *Prop.* 35, the point R is on the circle which passes through P, Q, Q'.

COR. 1. Since TP and TP' are equally inclined to the axis, the lines QQ', PR, which are parallel respectively to TP and TP', are also equally inclined to the axis.

COR. 2. When the point V is brought indefinitely near to P, QQ' coincides with the tangent to the parabola at P, and becomes also a tangent to the circle at P, since Q and Q' are indefinitely near to each other. The circle therefore becomes the circle of curvature at the point P.

Hence if PR be drawn parallel to the tangent at P', or be equally inclined to the axis with PT, it will meet the parabola in the point where the circle of curvature at P intersects the parabola.

PROBLEMS ON THE PARABOLA.

1. The diameter of the circle described about the triangle BAC is equal to $5AS$. (*See fig. Prop.* III.)

2. If from the point G, GK be drawn at right angles to SP, then $PK = 2AS$. (*See fig. Prop.* VII.)

3. If the triangle SPG is equilateral, then SP is equal to the latus rectum. (*See fig. Prop.* VII.)

4. PQ is a common tangent to a parabola and the circle described on the latus rectum as diameter; prove that SP and SQ make equal angles with the latus rectum.

5. Prove that $PY \cdot PZ = SP^2$, and that $PY \cdot YZ = AS \cdot SP$. (*See fig. Prop.* VII.)

6. If PL be drawn at right angles to AP, meeting the axis in L, and PN be the ordinate of P, then $NL = 4AS$.

7. The tangent at any point P of a parabola meets the directrix and latus rectum produced in points equally distant from the focus.

8. Prove that $NY = TY$, and that $TP \cdot TY = TS \cdot TN$. (*See fig. Prop.* VII.)

9. If a circle be described about the triangle SPN, the tangent to it from $A = \frac{1}{2} PN$. (*See fig. Prop.* VII.)

10. If the ordinate of a point P bisect the subnormal of P', the ordinate of P is equal to the normal of P'.

11. If from any point on the tangent to a parabola a line be drawn touching the parabola, the angle between this line and the line to the focus from the same point is constant.

12. A circle and parabola have the same vertex and axis. $BA'C$ is the double ordinate of the parabola which touches the circle at A', the extremity of the diameter through the vertex A. PP' is any other ordinate of the parabola parallel to this, meeting the axis in N, and AB produced in R; prove that the rectangle $RP \cdot RP'$ is proportional to the square of the tangent drawn from N to the circle.

13. Draw a parabola to touch a given circle at a given point, and such that its axis may touch the same circle in another given point.

14. If from the point of contact of a tangent to a parabola a chord be drawn, and another line be drawn parallel to the axis meeting the chord, tangent, and curve, this line will be divided by them in the same ratio as it divides the chord.

15. If the diameter PV meet the directrix in O, and the chord drawn through the focus parallel to the tangent at P in V, prove that $VP = PO$.

16. Prove that the locus of the intersection of a diameter PV with the chord drawn through the focus parallel to the tangent at P is a parabola.

17. If a circle and parabola have a common tangent at P, and intersect in Q and R; and QV, UR be drawn parallel to the axis of the parabola meeting the circle in V and U respectively, then VU is parallel to the tangent at P.

18. AB and AC are two lines at right angles to each other. From a fixed point C on AC, CR is drawn parallel to AB. On AR, produced if necessary, P is taken such that the perpendicular PN upon AB is equal to CR. Prove that the curve traced out by P is a parabola.

19. If from a point P of a circle PC be drawn to the centre; and R be the middle point of the chord PQ drawn parallel to a fixed diameter ACB, then the curve traced out by the intersection of CP and AR is a parabola.

20. If two equal tangents OQ, OQ', be cut by a third tangent, their alternate segments are equal.

21. E is the centre of the circle described about the triangle OQQ'. Prove that the circle described about the triangle QEQ' will pass through the focus. (*See fig. Prop.* XIII.)

22. PSp is any focal chord of a parabola. Prove that AP, Ap will meet the latus rectum in two points Q, q, whose distances from the focus are equal to the ordinates of p and P.

23. PSp is a focal chord of a parabola, RDr the directrix meeting the axis in D; and Q any point on the curve. Prove that if QP, Qp be produced to meet the directrix in R, r, half the latus rectum is a mean proportional between DR, Dr.

24. OP and OQ are two tangents to a parabola. On QO produced, OQ' is taken equal to OQ; prove that $OS \cdot PQ' = OP \cdot OQ$.

25. If QD be drawn at right angles to the diameter PV, then $QD^2 = 4AS \cdot PV$.

26. If through any point O on the axis of a parabola a chord POQ be drawn, and PM, QN be the ordinates of the points P and Q, prove that $AM \cdot AN = AO^2$.

27. If AP and AQ be drawn at right angles to each other from the vertex of a parabola, and PM, QN be the ordinates of P and Q, prove that the latus rectum is a mean proportional between AM and AN.

28. OAP is the sector of a circle whose centre is O. If the radius OA remain fixed while the angle AOP changes the centre of the circle inscribed in the sector, AOP will trace out a parabola.

29. QSQ' is a focal chord parallel to AP; PN, QM, $Q'M'$ are the ordinates of P, Q, and Q'. Prove that $SM^2 = AM \cdot AN$ and that $MM' = AP$.

30. PQ, PQ' are drawn from any point P cutting the ordinates $Q'V'$, QV in R' and R, prove that VR is to $V'R'$ in the triplicate ratio of QV to $Q'V'$.

31. On a chord of a parabola as diameter a circle is described cutting the parabola again in two points. If these points be joined, the portion of the axis between the two chords is equal to the latus rectum.

32. If OQ, OQ' be a pair of tangents to a parabola, and the chord QQ' be a normal to the curve at Q, then OQ is bisected by the directrix.

33. Two equal parabolas having the same focus and their axes in contrary directions intersect at right angles.

34. The radius of curvature at the extremity of the latus rectum is equal to twice the normal.

35. If from any point P of a parabola PF and PH be drawn making equal angles with the normal PG, then $SG^2 = SF \cdot SH$.

36. If a triangle be inscribed in a parabola, the points when the sides produced meet the tangents at the opposite angles are in the same straight line.

37. If the tangents OQ, OQ' be cut by a third tangent in R, R', prove that
$$OR : RQ :: R'Q' : OR'.$$

38. If from the vertex of a parabola chords be drawn at right angles to one another, and on them a rectangle be described, the curve traced out by the further angle is a parabola.

39. Prove that $2PY$ is a mean proportional between AP and the chord of the circle of curvature at the point P of the parabola drawn through the vertex A. (*See fig. Prop.* VII.)

40. If a circle described upon the chord of a parabola as diameter meet the directrix it also touches it, and all chords for which this is possible intersect in a point.

41. If a parabola *roll* upon another equal parabola, the vertices originally coinciding, the focus traces out the directrix.

42. The circle of curvature at the extremity of the latus rectum intersects the parabola on the diameter of curvature passing through the point of contact.

CHAPTER II.

THE ELLIPSE.

22. DEF. The *Ellipse* is the curve traced out by a point which moves in such a manner that its distance from a given fixed point continually bears the same ratio, *less than unity*, to its distance from a given fixed line. (*See Introduction.*)

PROP. I.

The focus and directrix of an ellipse being given, to find any number of points on the curve.

Let S be the focus, and MX the directrix.

Draw SX at right angles to the directrix, and divide SX in the point A, so that SA may be to AX in the given fixed ratio less than unity ; then

A is a point on the curve.

On XS produced take a point A', such that

$$SA' : A'X :: SA : AX;$$

then A' will also be a point on the curve.

On the directrix take *any* point M; and through M and S draw the line $MYSY'$ meeting AY and $A'Y'$, drawn at right angles to AA' in the points Y and Y'.

On YY' as diameter describe a circle, and draw MPP' parallel to AA', cutting the circle in the points P and P';

P and P' will be points on the ellipse.

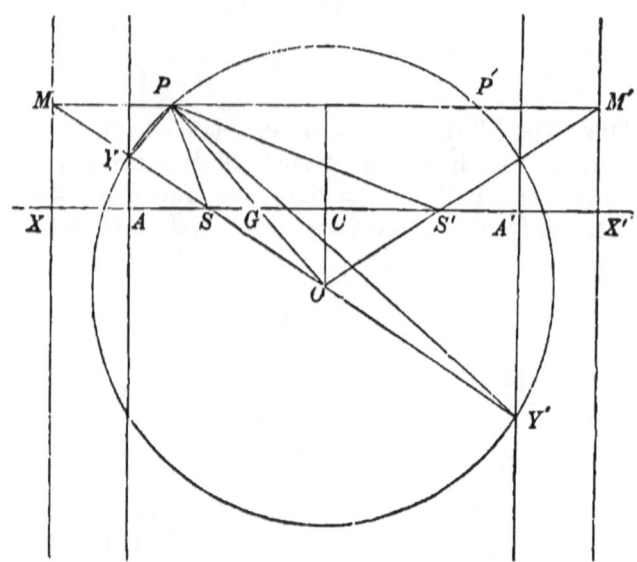

Join PY, PY', SP; then since
$$SY : YM :: SA : AX, \quad (Euclid, \text{VI. 2.})$$
and $SY' : Y'M :: SA' : A'X,\quad (Euclid, \text{VI. 2.})$
$$\therefore SY : YM :: SY' : Y'M;$$
or, alternately,
$$SY : SY' :: YM : Y'M,$$
and the angle YPY' in a semicircle is a right angle,
$$\therefore PY \text{ bisects the angle } SPM,^*$$
$$\therefore SP : PM :: SY : YM,$$
$$:: SA : AX.$$
So we may show that
$$SP' : P'M :: SY : YM,$$
$$:: SA : AX,$$
$\therefore P$ and P' are points on the curve.

---

* For, if not, make the angle YPs equal to YPM; then
$$sY : YM :: sP : PM. \quad (Euclid, \text{VI. 3.})$$
And since, if PY bisect sPM, PY', being at right angles to PY, also bisects the angle sPM',
$$\therefore sY' : Y'M :: sP : PM. \quad (Euclid, \text{VI. A.})$$
Hence $sY : YM :: sY' : Y'M,$
or $sY : sY' :: YM : Y'M,$
$\therefore$ the points S and s coincide.

In the same way, by taking other points on the directrix, we may obtain as many more points on the curve as we please.

Cor. 1. Since, corresponding to every point P on the curve, there is a point P' situated in precisely the same manner with respect to $A'Y'$ as P is with respect to AY, it is clear that if we make $A'S'$ equal to AS, and $A'X'$ equal to AX, and draw $X'M'$ at right angles to AX', the curve could be equally well described with S' as focus and $M'X'$ as directrix.

The ellipse is therefore symmetrical, not only with respect to the line AA', but also with respect to the line OC drawn through the middle point of YY' at right angles to and bisecting AA'.

Cor. 2. The line OP will bisect the angle SPS'.

Let OP meet SS' in G. Produce MP to meet $X'M'$ in M', and draw OM' passing through the focus S'; then

$$SP : PM :: S'P : PM',$$

or, alternately, $SP : S'P :: PM : PM'$. (1)

Again, $SG : PM :: S'G : PM'$,

or, alternately, $SG : S'G :: PM : PM'$, (2)

$\therefore$ from (1) and (2)

$$SP : S'P :: SG : S'G,$$

$\therefore PG$ bisects the angle SPS'. (*Euclid*, VI. 3.)

It will be shown hereafter (*Prop.* XI.) that the normal to the ellipse at the point P also bisects the angle SPS'. Hence the ellipse and circle have the same tangent at the point P.

The ellipse will consequently touch all the infinite series of circles which can be described in the same manner as the one in the figure by taking different points on the directrix.

Prop. II.

23. If C be the middle point of AA', then CA is a mean proportional between CS and CX,

or $CS \cdot CX = CA^2$. (*See fig. Prop.* III.)

Since $SA' : A'X :: SA : AX$.

Alternately $SA' : SA :: A'X : AX$,

$\therefore SA' + SA : SA :: A'X + AX : AX$;

or $AA' : SA :: XX' : AX$,

$\therefore AA' : XX' :: SA : AX$,

or $CA : CX :: SA : AX$. (1)*

Again, $SA' : SA :: A'X : AX$,

$\therefore SA' - SA : SA :: A'X - AX : AX$;

or $SS' : SA :: AA' : AX$.

Alternately $SS' : AA' :: SA : AX$;

or $CS : CA :: SA : AX$. (2)

Hence from (1) and (2)

$CA : CX :: CS : CA$,

$\therefore CA^2 = CX \cdot CS$;

or CA is a mean proportional between CS and CX.

Cor. Since the three lines CS, CA, CX are proportional, therefore, by the definition of duplicate ratio and *Euclid*, VI. 20 *Cor.*

$CS : CX : CS^2 : CA^2$. (3)

Prop. III.

24. If P be any point on the ellipse, then
$$SP + S'P = AA'.$$

---

* N.B. The results (1), (2), (3) should be remembered, as they will frequently be referred to.

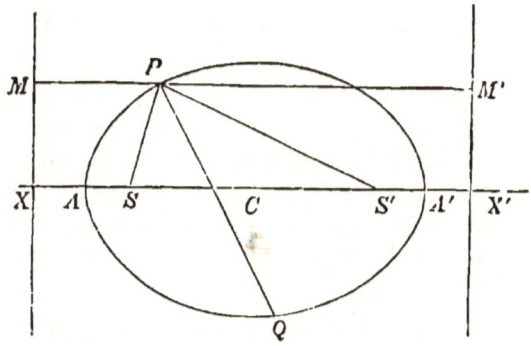

Since $SP : PM :: SA : AX$,
and $SA : AX :: AA' : XX'$, (*Prop.* II.)
$\therefore SP : PM :: AA' : XX'$,
So $S'P : PM' :: AA' : XX'$,
$\therefore SP + S'P : PM + PM' :: AA' : XX'$.
But $PM + PM' = MM' = XX'$,
$\therefore SP + S'P = AA'$.

Cor. 1. By means of this property the ellipse may be practically described and the form of the curve determined.

Let a string, equal in length to AA', have its ends fastened to two points S and S'; and let it be kept stretched by means of the point of a pencil at P; then since $SP + S'P$ will be always equal to AA', the point P will trace out the ellipse.

Cor. 2. The line AA' is the longest line that can be drawn in the ellipse.

For, if any other line PQ be drawn, then
$$SP + SQ > PQ,$$
$$\text{and } S'P + S'Q > PQ,$$
$$\therefore SP + S'P + SQ + S'Q > 2PQ,$$
$$\text{or } AA' > PQ.$$

25. Def. If BCB' be drawn at right angles to ACA', meeting the ellipse in B and B', it will be seen further on (*Prop.* XIII. *Cor.* 2) that BCB' is the shortest chord that can be drawn through the *centre* of the ellipse. (See *fig. Prop.* IV.)

AA' is called the *Major Axis* and BB' the *Minor Axis* of the ellipse.

In most geometrical treatises the ellipse is defined as the curve traced out by a point which moves in such a manner that the sum of its distances from two fixed points is always the same; but it appears that the properties of the curve are more clearly exhibited by defining it in a manner analogous to the parabola, and deducing *immediately* from that definition the property in question.

Having now shown that one definition necessarily includes the other, we are at liberty in our future investigations to make use of whichever property is most convenient.

PROP. IV.

26. If BC be the semi-minor axis of the ellipse, then
$$BC^2 = CA^2 - CS^2;$$
and if SL be the semi-latus rectum,
$$SL \cdot AC = BC^2.$$

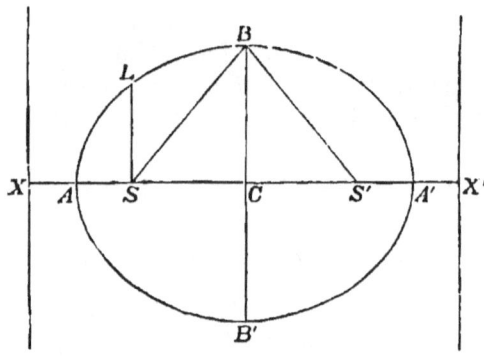

Join SB, $S'B$; then
since $SB + S'B = AA'$, (*Prop.* III.)
and that $SB = S'B$,
∴ $SB = AC$.
But $BC^2 = SB^2 - CS^2$,
∴ $BC^2 = CA^2 - CS^2$,

Again, $\quad SL : SX :: SA : AX$,
$$:: CS : CA, \quad (Prop.\ II.)$$
$\therefore SL \cdot AC = CS \cdot SX$,
$$= CS \cdot CX - CS^2, \quad (Euclid,\ II.\ 3,)$$
$$= CA^2 - CS^2, \quad (Prop.\ II.)$$
$$= BC^2.$$

Prop. V.

27. The sum of the distances of *any* point from the foci of an ellipse will be less or greater than AA' according as the point is inside or outside the ellipse.

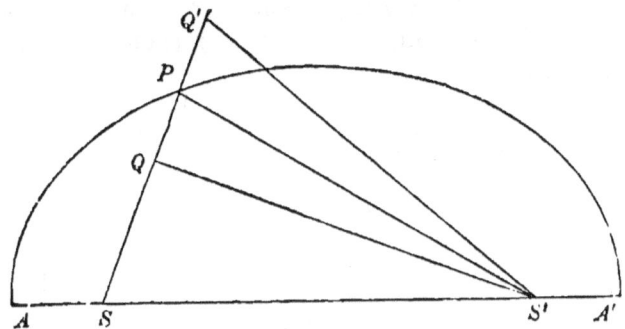

(1) Let Q be a point inside the ellipse.

Join $SQ,\ S'Q$; and produce SQ to meet the ellipse in P; join $S'P$; then
$$\text{since } S'P + QP > S'Q,$$
$$\therefore S'P + SP > S'Q + SQ.$$
But $S'P + SP = AA'$, $\quad (Prop.\ III.)$
$$\therefore SQ + SQ' < AA'.$$

(2) Let Q' be a point outside the ellipse.

Join $SQ,\ S'Q$, and let SQ meet the ellipse in the point P: join $S'P$; then
$$\text{since } S'Q + QP > S'P,$$
$$\therefore S'Q + SQ > SP + S'P,$$
But $SP + S'P = AA'$, $\quad (Prop.\ III.)$
$$\therefore SQ + S'Q > AA'.$$

36 CONIC SECTIONS.

Cor. Conversely, a point will be inside or outside the ellipse, according as the sum of its distances from the foci is less or greater than AA'.

28. Def. If a point P' be taken on the ellipse near to P, (see fig. Prop. VI.) and PP' be joined, the line PP' produced, in the limiting position which it assumes when P' is made to approach indefinitely near to P, is called the *Tangent* to the ellipse at the point P.

Prop. VI.

If the tangent to the ellipse at any point P intersect the directrix in the point Z, and if S be the focus corresponding to the directrix on which Z is situated, then SZ will be at right angles to SP.

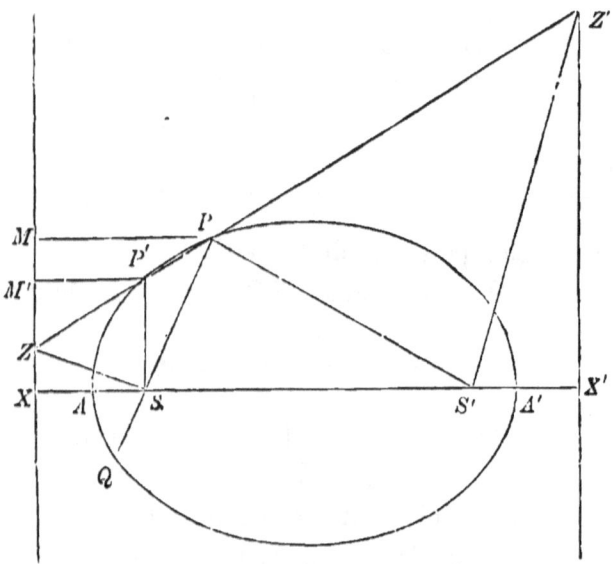

Let P' be a point on the ellipse near to P.

Draw the chord PP', and produce it to meet the directrix in Z; join SZ.

Draw PM, $P'M'$ at right angles at the directrix, and join SP, SP'.

Produce PS to meet the ellipse in the point Q; then since the triangles ZMP, $ZM'P'$ are similar,

$$\therefore ZP : ZP' :: MP : M'P',$$
$$:: SP : SP',$$

$\therefore SZ$ bisects the angle $P'SQ$. (*Euclid*, VI. *Prop.* A.)

Now when P' is indefinitely near to P, and PP' becomes the tangent at the point P, the angle PSP' becomes indefinitely small, while the angle QSP' approaches two right angles; and therefore the angle ZSP' being half of the angle $P'SQ$, becomes ultimately a right angle.

Hence when PZ becomes the tangent at the point P,
the angle ZSP is a right angle,
or SZ is perpendicular to SP.

COR. 1. Conversely, if SZ be drawn at right angles to SP, meeting the directrix in Z, and PZ be joined, PZ will be the tangent at P.

COR. 2. If ZP be produced to meet the other directrix on the point Z' and $S'Z'$ be joined, then
$S'Z'$ is at right angles to $S'P$.

COR. 3. The tangents at the extremities of the latus rectum or double ordinate through the focus meet the axis produced in the point X.

PROP. VII.

The tangent to the ellipse at any point P makes equal angles with the focal distances SP and $S'P$.

Let the tangent at P meet the directrices in Z and Z'.

Draw MPM' at right angles to the directrices, meeting them in M and M' respectively; join SZ, $S'Z'$; then
$$SP : PM :: S'P : PM';$$
and since the triangles MPZ, $M'PZ'$, are similar,
$$PM : PZ :: PM' : PZ',$$
$$\therefore SP : PZ :: S'P : PZ'. \quad (\textit{Ex æquali.})$$

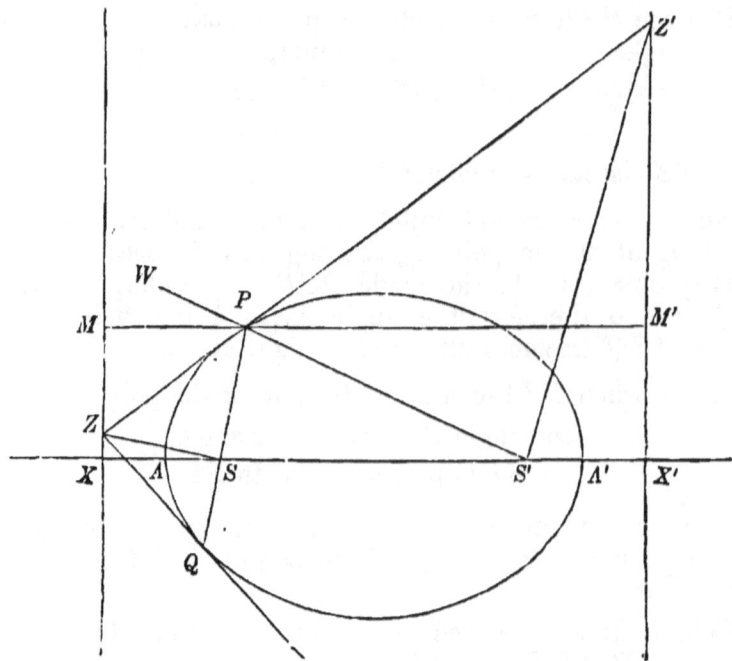

Now in the triangles SPZ, $S'PZ'$, because the sides about the angles SPZ, $S'PZ'$ are proportional, and the angles PSZ, $PS'Z'$ are equal, being right angles, and the angles SZP, $S'Z'P$ are each less than a right angle,

∴ the triangles SPZ and $S'PZ'$ are similar, (*Euclid*, VI. 7)

∴ the angle SPZ = the angle $S'PZ'$.

Cor. If $S'P$ be produced to W; then

the angle SPZ = the angle WPZ.

Prop. VIII.

The tangents at the extremities of a focal chord intersect in the directrix.

Let PSQ be a focal chord, and let the tangent at P meet the directrix in Z.

Join SZ; then

the angle ZSP is a right angle, (*Prop.* VI.)
and $\therefore$ also the angle ZSQ is a right angle,
$\therefore ZQ$ is the tangent at Q; (*Prop.* VI. *Cor.* 1)
or the tangents at the extremities of a focal chord intersect in the directrix.

Prop. IX.

29. If the tangent at P meet the axis major produced in T, and PN be the ordinate of the point P, then
$$CT \cdot CN = CA^2.$$

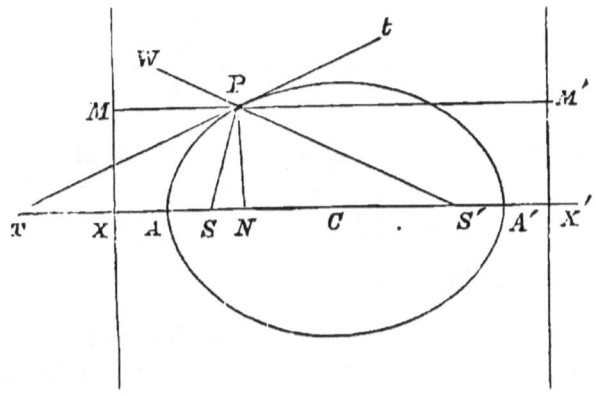

Draw MPM' parallel to the axis major meeting the directrices in M and M'; and produce $S'P$ to W; then, since PT bisects the angle SPW, (*Prop.* VII. *Cor.*)

$\therefore S'T : ST :: S'P : SP$, (*Euclid*, VI. A.)
$:: PM' : PM$,
$:: X'N : XN$,
$\therefore S'T + ST : S'T - ST :: X'N + XN : X'N - XN$;
or $2CT : 2CS :: 2CX : 2CN$,
or $CT : CS :: CX : CN$,
$\therefore CT \cdot CN = CS \cdot CX$,
$= CA^2$. (*Prop.* II.)

Prop. X.

If on the major axis of an ellipse as diameter a circle be described and a common ordinate NPQ be drawn meeting the ellipse in P and the circle in Q, then the tangents to the ellipse and circle respectively at the points P and Q will meet the major axis produced in the same point.

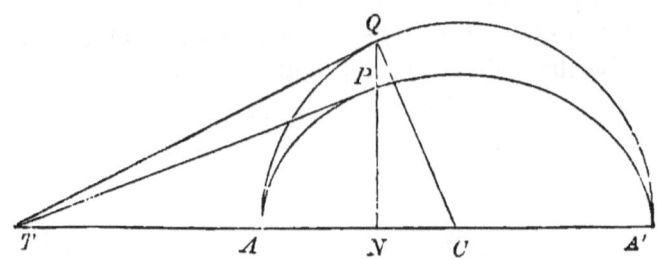

Let the tangent to the ellipse at P meet the major axis produced in T; join CQ, QT; then, by the last Proposition,

$$CT \cdot CN = CA^2 = CQ^2,$$

∴ the angle CQT is a right angle.

And therefore QT is the tangent to the circle at Q; or the tangents at P and Q meet the major axis produced in the same point T.

The circle described on AA' as diameter is called the *Auxiliary Circle* on account of the important aid that it affords in investigating the properties of the ellipse.

30. DEF. The line PG, drawn at right angles to the tangent PT, is called the *Normal* to the ellipse at the point P.

Prop. XI.

If the normal at P meet the major axis in the point G; then

$$SG : SP :: CS : CA,$$

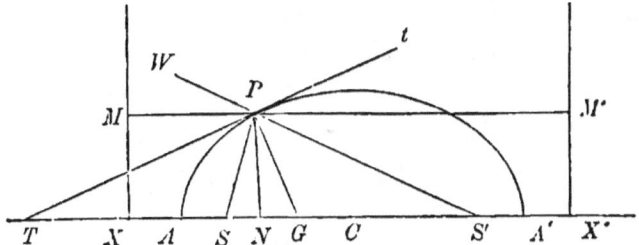

Since PG is at right angles to TPt,

∴ the angle GPT = the angle GPt.

But the angle SPT = the angle $S'Pt$, (*Prop.* VII.)

∴ the angle SPG = the angle $S'PG$,

or PG bisects the angle SPS',

∴ $SG : S'G :: SP : S'P$, (*Euclid*, VI. 3.)

∴ $SG : SG + S'G :: SP : SP + S'P$;

or $SG : SS' :: SP : AA'$;

or $SG : SP :: SS' : AA'$;

or $SG : SP :: CS : CA$.

Cor. Hence also,

$$S'G : S'P :: CS : CA.$$

Prop. XII.

31. If the normal at P meet the major axis in G, and PN be the ordinate at the point P, then (*see fig. Prop.* XI.)

$$NG : NC :: BC^2 : AC^2.$$

Draw MPM' parallel to the axis meeting the directrices in M and M'; join SP, $S'P$; then, since PG bisects the angle SPS', (*Prop.* XI.)

∴ $S'G : SG :: S'P : SP$,

:: $PM' : PM$,

:: $X'N : XN$,

∴ $S'G - SG : S'G + SG :: X'N - XN : X'N + XN$;

or $2CG : SS' :: 2CN : XX'$.

Alternately, $2CG : 2CN :: SS' : XX'$;

or $CG : CN :: CS : CX$,

$:: CS^2 : CA^2$, (*Prop.* II. *Cor.*)

$\therefore CN - CG : CN :: CA^2 - CS^2 : CA^2$;

or $NG : CN :: BC^2 : AC^2$.

Prop. XIII.

32. If PN be the ordinate of any point P on the ellipse; then

$$PN^2 : AN \cdot A'N :: BC^2 : AC^2.$$

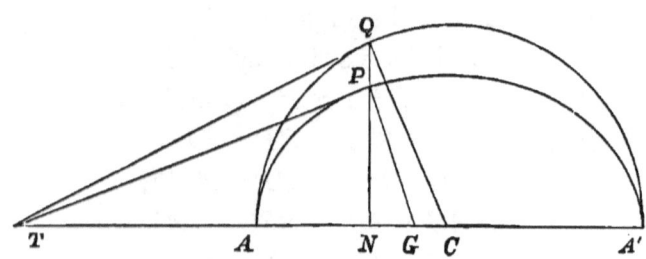

Produce NP to meet the auxiliary circle in the point Q, and draw the tangents PT, QT meeting the major axis produced in the point T. (*Prop.* X.)

Join CQ, and let the normal at P meet the ellipse in G then, by the last Proposition,

$$NG : CN :: BC^2 : AC^2.$$

And rectangles of the same altitude are to another as their bases,

$\therefore TN \cdot NG : TN \cdot CN :: BC^2 : AC^2$;

or $PN^2 : QN^2 :: BC^2 : AC^2$. (*Euclid*, VI. 8, *Cor.*)

But $QN^2 = AN \cdot A'N$,

since the angle AQA' in a semicircle is a right angle,

$\therefore PN^2 : AN \cdot A'N :: BC^2 : AC^2$.

Cor. 1. Also $PN : QN :: BC : AC$.

This result is the basis of many of the future Propositions of the ellipse.

Cor. 2. Since $PN^2 : QN^2 :: BC^2 : AC^2$,

$\therefore PN^2 : AC^2 - CN^2 :: BC^2 : AC^2$,

$\therefore PN^2 : AC^2 - CN^2 - PN^2 :: BC^2 : AC^2 - BC^2$,

or $PN^2 : AC^2 - CP^2 :: BC^2 : AC^2 - BC^2$.

Now PN^2 is always less than BC^2,

$\therefore CP^2$ is greater than BC^2,

$\therefore BC$ is the shortest line that can be drawn to the ellipse from the centre.

Prop. XIV.

If the tangent at any point P of an ellipse meet the minor axis CB produced in t, and Pn be drawn at right angles to CB; then

$$Ct \cdot Cn = BC^2,$$

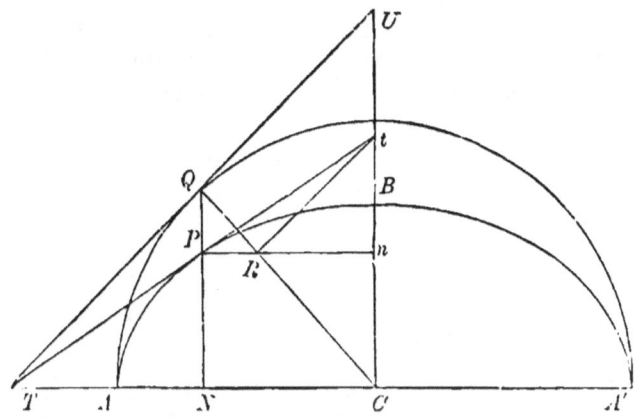

Draw the common ordinate NPQ to the ellipse and the auxiliary circle; and let the tangents at P and Q to the ellipse and circle respectively meet the major axis produced in T (*Prop.* X.) and the minor axis produced in t and U.

Join CQ meeting Pn in R; then since PR is parallel to CN,

$CR : CQ :: PN : QN$.

$:: BC : AC$. (*Prop.* XIII. *Cor.* 1.)

But $CQ = AC$,

$\therefore CR = BC$. (1)

Again, joining Rt,

$$Ct : CU :: PN : QN,$$
$$:: CR : CQ,$$

$\therefore Rt$ is parallel to QU,

$\therefore CRt$ is a right angle,

$\therefore Ct \cdot Cn = CR^2$ (*Euclid*, VI. 8, *Cor.*)

But $CR = BC$, (1)

$\therefore Ct \cdot Cn = BC^2$.

This proposition also admits of a demonstration similar to that given for the corresponding property of the hyperbola.

PROP. XV.

33. If from the foci S and S', SY and $S'Y'$ are drawn at right angles to the tangent at P, then Y and Y' are on the circumference of the auxiliary circle, and

$$SY \cdot S'Y' = BC^2.$$

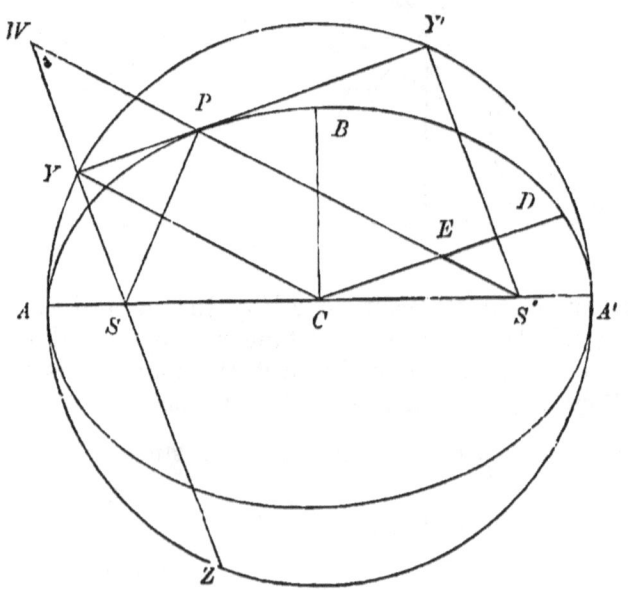

Join SP, $S'P$, and produce SY and $S'P$ to meet in W; and join CY; then

since the angle SPY = the angle WPY, (*Prop.* VII. *Cor.*)

and the angle SYP = the angle WYP,

and the side PY is common to the triangles SPY, WPY,

$\therefore$ the triangle SPY = the triangle WPY in all respects,

$$\therefore SP = PW,$$
$$\therefore SP + S'P = S'W.$$
But $SP + S'P = AA'$, (*Prop.* III.)
$$\therefore S'W = AA'.$$

Again $\because SC = S'C$ and $SY = YW$,
$$\therefore SC : S'C :: SY : YW,$$
$$\therefore CY \text{ is parallel to } S'W,$$
$$\therefore CY : S'W :: CS : SS',$$
$$\therefore CY = \tfrac{1}{2} S'W = CA.$$
So $CY' = CA$,

$\therefore Y$ and Y' are points on the auxiliary circle.

Next let YS be produced to meet the auxiliary circle in Z, and join ZY'; then

since the angle ZYY' is a right angle,

$\therefore ZY'$ passes through the centre C,

$\therefore$ the angle SCZ = the angle $S'CY'$.

$$\therefore SZ = S'Y',$$
$$\therefore SY \cdot S'Y' = SY \cdot SZ,$$
$$= AS \cdot A'S, \quad (Euclid, \text{ III. } 35)$$
$$= CA^2 - CS^2, \quad (Euclid, \text{ II. } 5)$$
$$= BC^2. \quad (Prop. \text{ IV.})$$

Cor. If CD be drawn parallel to the tangent at P, meeting SP in E; then

since the figure $CYPE$ is a parallelogram,
$$\therefore PE = CY = AC.$$

Prop. XVI.

34. To draw a pair of tangents to an ellipse from an external point O.

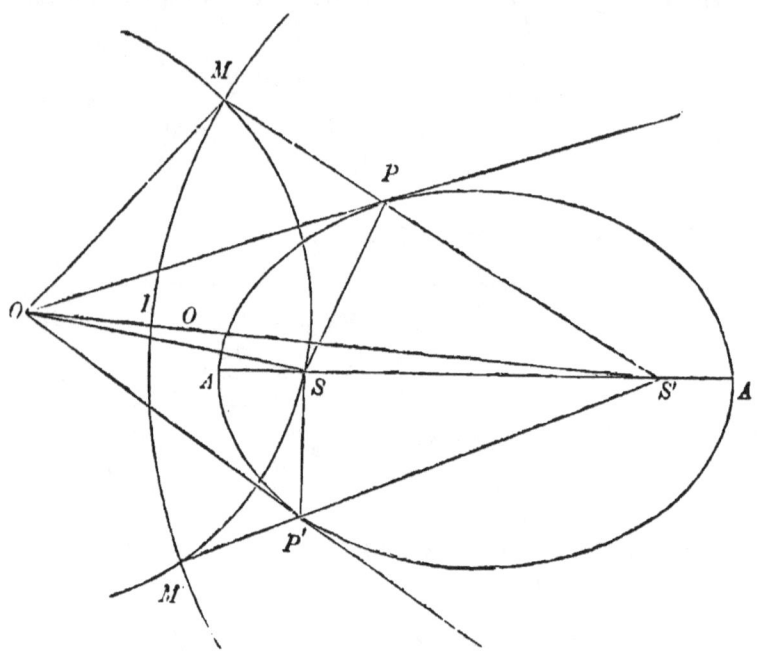

With centre S' and radius equal to AA' describe a circle.

Join OS, OS'; and let SO or $S'O$ produced meet the circle in the point I.

Now, if O be a point outside the circle MIM', it is evident that OS is greater than OI; and if O be inside the circle,
$$\text{since } OS + OS' > AA' \text{ or } S'I, \quad (Prop.\ V.)$$
$$\therefore OS > OI.$$

With centre O and radius OS describe another circle cutting the former in the points M and M', which it will always do, since OS is greater than OI.

Join $S'M$, $S'M'$, meeting the ellipse in the points P and P'.
Join OP, OP'; these will be the tangents required.
Join SP, SP'; then, since
$$SP + S'P = AA' = S'M,$$
$$\therefore SP = PM.$$
And $\because SP$, $PO = MP$, PO, each to each,
and $OS = OM$,
$\therefore$ the angle OPS = the angle OPM,
$\therefore OP$ is the tangent at P. (*Prop.* VII. *Cor.*)
So OP' is the tangent at P'.

Prop. XVII.

If from a point O a pair of tangents OP, OP' be drawn to an ellipse, then OP and OP' will subtend equal angles at either focus.

Join SP, $S'P$; SP', $S'P'$; and produce $S'P$, $S'P'$ to M and M', making PM equal to SP, and $P'M'$ equal to SP'.
Join OM, OM'; OS, OS'.
Then since OP, $PS = OP$, PM, each to each,
and the angle OPS = the angle OPM, (*Prop.* VII. *Cor.*)
$$\therefore OS = OM,$$
and the angle OSP = the angle OMP.
So $OS = OM'$,
and the angle OSP' = the angle $OM'P'$,
$$\therefore OM = OM'.$$
Again, $\because S'M = S'P + SP = AA'$,
and $S'M' = S'P' + SP' = AA'$,
$$\therefore S'M = S'M',$$
And $\because OS'$, $S'M = OS'$, $S'M'$, each to each,
and $OM = OM'$,
$\therefore$ the angle $OS'M$ = the angle $OS'M'$,
and the angle OMS' = the angle $OM'S'$.

But the angle OMS' = the angle OSP,
and the angle $OM'S'$ = the angle OSP',
∴ the angle OSP = the angle OSP',
∴ OP and OP' subtend equal angles at either focus.

Prop. XVIII.

35. If from an external point O a pair of tangents OQ, OQ' be drawn to an ellipse, and CO be joined meeting the chord QQ' in V, and the ellipse in P; then

(1.) QQ' will be bisected in V.

(2.) The tangent at P will be parallel to QQ'.

(3.) CP will be a mean proportional between CV and CO.

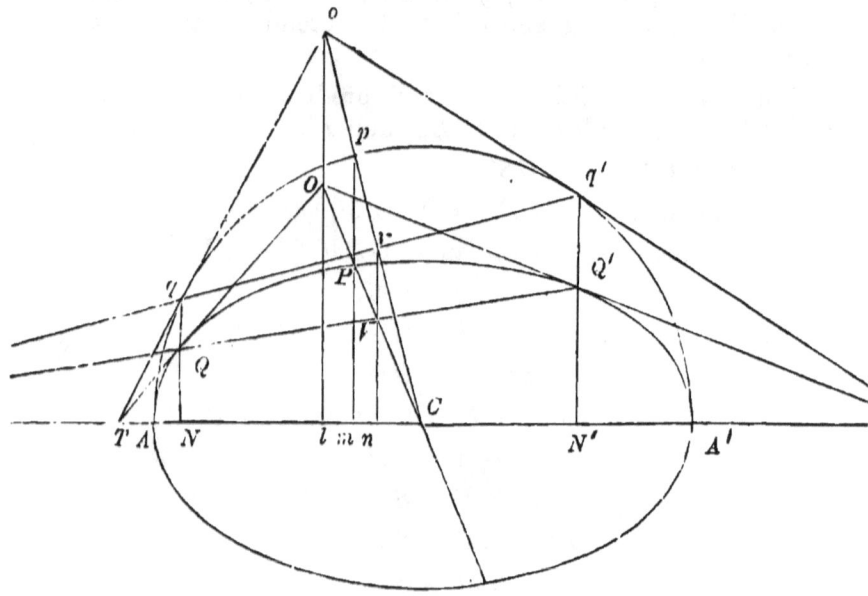

Produce OQ, OQ' to meet the major axis produced in T' and T''.

Draw the ordinates NQ, $N'Q'$, and produce them to meet the circle in q and q'.

Then Tq and $T'q'$ will be tangents to the auxiliary circle. (*Prop.* X.)

Let Tq and $T''q'$ be produced to meet in o; join Co meeting the chord qq' in v, and the circle in p.

Now, since the corresponding ordinates of the ellipse and auxiliary circle are in the constant ratio of BC to AC, the three lines ol, pm, vn drawn at right angles to AA' will pass through the points O, P, V respectively.

For, according as O is the point where ol meets TQ or $T''Q'$ we shall have

$$lO : lo :: NQ : Nq,$$
$$:: BC : AC;$$
or $lO : lo :: N'Q' : N'q',$
$$:: BC : AC,$$

$\therefore Oo$ is perpendicular to AA'.

So Pp and Vv are perpendicular to AA',

$\therefore Oo, Pp, Vv$ are parallel.

Hence (1.) $QV : VQ' :: qv : vq'$.

But $qv = vq'$ from the circle,

$\therefore QV = VQ'$;

or QQ' is bisected in V.

(2.) Since $NQ : Nq :: N'Q' : N'q'$

it is evident that QQ' and qq' will meet the axis produced in the same point.

Also the tangents to the ellipse and circle at P and p respectively will meet the axis in the same point.

Now in the circle the tangent at p is manifestly parallel to qq',

and $NQ : Nq :: mP : mp$,

$\therefore$ the tangent at P is parallel to QQ'.

(3.) If Cq be joined, since the angle Cqo is a right angle and Co is perpendicular to qq',

$\therefore Cv : Cq :: Cq : Co,$ (*Euclid*, VI. 8 *Cor.*)

or, since $Cq = Cp$,

$Cv : Cp :: Cp : Co$.

E

But $Cv : Cp :: CV : CP$,
and $Cp : Co :: CP : CO$,
∴ $CV : CP :: CP : CO$,
∴ $CO . CV = CP^2$.

Cor. From this it is manifest that if any number of chords be drawn parallel to each other in an ellipse, their middle points will all lie on the line drawn from the centre to the point where the tangent parallel to the chord meets the ellipse.

Def. The line PCP' drawn through the centre of an ellipse and meeting the curve in P and P', is called a *Diameter*.

The diameter consequently bisects all chords parallel to the tangents at its extremities; and the tangents at the extremities of any chord will intersect the diameter corresponding to that chord in the same point.

36. Def. If CD be drawn parallel to the tangent at P, then CD is said to be *conjugate* to CP.

Prop. XIX.

In the ellipse if CD be conjugate to CP, then will CP be conjugate to CD.

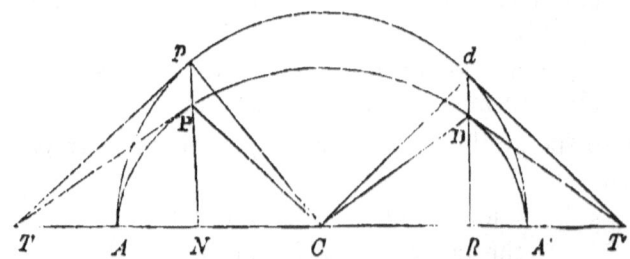

Draw the ordinates PN, DR, and produce them to meet the auxiliary circle in the points p, d.

Join CP, Cp; CD, Cd; and draw the tangents TP, Tp; $T'D$, $T'd$.

Now, since CD is parallel to PT,

∴ the triangle PNT is similar to the triangle DRC.

∴ $TN : CR :: PN : DR$,

∴ $Np : Rd$, (Prop. XIII. Cor.)

∴ Tp is parallel to Cd,

∴ the angle pCd is a right angle,

∴ Cp is parallel to $T''d$,

∴ the triangle pCN is similar to the triangle $dT''R$,

∴ $NC : RT'' :: Np : Rd$,

$:: NP : RD$,

∴ CP is parallel to DT',

∴ CP is conjugate to CD.

Cor. Since CRd and CNp are each similar to dRT', (*Euclid*, VI. 8)

∴ the triangle CRd is similar to the triangle CNp,

and the side Cd = the side Cp,

∴ the triangle CRd = the triangle CNp in all respects,

∴ $CN = Rd$, and $CR = Np$.

Hence $DR : CN :: DR : Rd$,

$:: BC : AC$;

also $PN : CR :: PN : Np$,

$:: BC : AC$.

Prop. XX.

37. If CP and CD be conjugate semi-diameters, and PN, DR be the ordinates of the points P and D; then

(1.) $CN^2 + CR^2 = AC^2$.

(2.) $PN^2 + DR^2 = BC^2$.

(3.) $CP^2 + CD^2 = AC^2 + BC^2$.

e 2

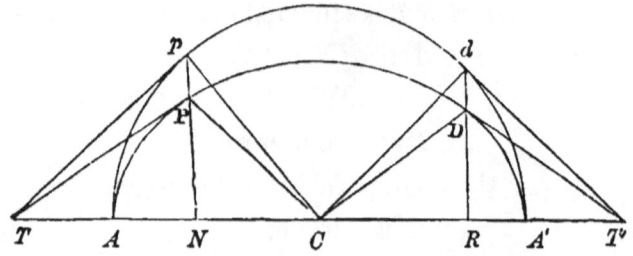

Produce NP, RD to meet the auxiliary circle in the points p, d; then

$$CN = Rd, \quad (Prop.\ XIX.\ Cor.)$$
$$\therefore CN^2 + CR^2 = Rd^2 + CR^2,$$
$$= Cd^2,$$
$$= CA^2.$$

Again, $PN : Np :: BC : AC$,
$$\therefore PN^2 : Np^2 :: BC^2 : AC^2.$$
So $DR^2 : Rd^2 :: BC^2 : AC^2$,
$$\therefore PN^2 + DR^2 : Np^2 + Rd^2 :: BC^2 : AC^2;$$
but $Np^2 + Rd^2 = CR^2 + CN^2$,
$$= AC^2,$$
$$\therefore PN^2 + DR^2 = BC^2,$$
and $CN^2 + CR^2 = AC^2$,
$$\therefore CP^2 + CD^2 = AC^2 + BC^2.$$

38. DEF. A line QV drawn parallel to the tangent at P, and meeting CP in V, is called an *Ordinate* to the diameter CP.

PROP. XXI.

If QV be any ordinate to the diameter PCP', and CD be conjugate to CP; then

$$QV^2 : PV \cdot P'V :: CD^2 : CP^2.$$

Draw the tangent UQW meeting CP and CD produced in U and W; and draw QR parallel to CP, meeting CD in R.

Now, since $CR : CD :: CD : CW$, (*Prop.* XVIII.)

$\therefore CR^2 : CD^2 :: CR : CW$, (*Euclid*, VI. 20 *Cor.*)

or $QV^2 : CD^2 :: UV : CU$.

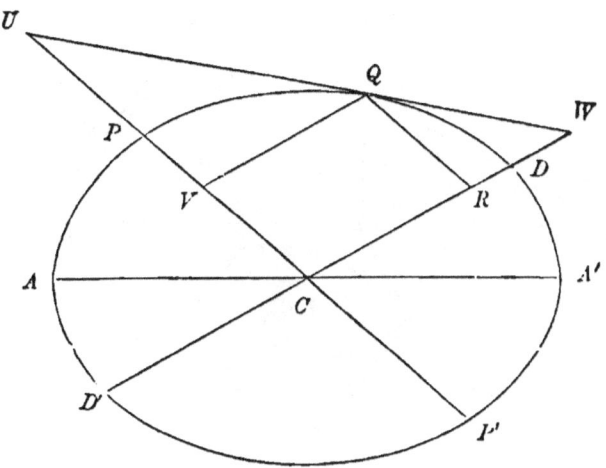

Again,

since $CU : CP :: CP : CV$, (*Prop.* XVIII.)

$\therefore CU : CV :: CP^2 : CV^2$, (*Euclid*, VI. 20 *Cor.*)

$\therefore CU - CV : CU :: CP^2 - CV^2 : CP^2$,

or $UV : CU :: PV \cdot P'V : CP^2$.

Hence $QV^2 : CD^2 :: PV \cdot P'V : CP^2$,

or $QV^2 : PV \cdot P'V :: CD^2 : CP^2$.

Prop. XXII.

39. The area of any parallelogram formed by drawing tangents to an ellipse at the extremities of a pair of conjugate

diameters is equal to the rectangle contained by the axes of the ellipse.

Let PCP', DCD' be a pair of conjugate diameters, and let a parallelogram be formed by drawing tangents at the points P, P', D, D'.

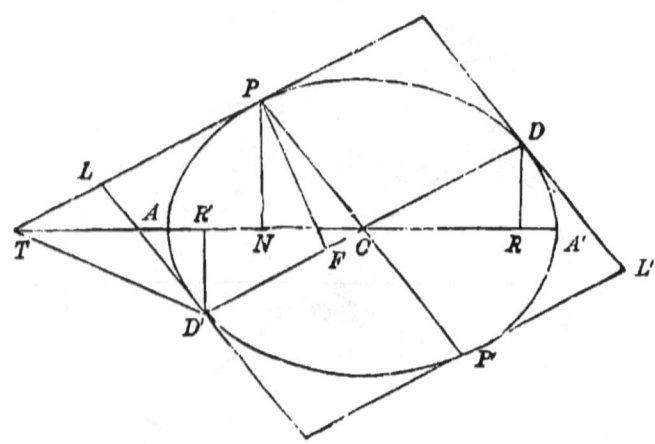

Let the tangent at P meet CA produced in T; join $D'T$.

Draw the ordinates PN, DR, $D'R'$; then since PT is parallel to CD', the parallelogram PD' is double the triangle CTD', and therefore equal to the rectangle contained by CT and $D'R'$.

Now $D'R' : CN :: BC : AC$, (*Prop.* XIX. *Cor.*)

∴ $CT \cdot D'R' : CT \cdot CN :: BC : AC$, (*Euclid*, VI. 1.)

:: $BC \cdot AC : AC^2$. (*Euclid*, VI. 1.)

But $CT \cdot CN = AC^2$,

∴ $CT \cdot D'R' = AC \cdot BC$,

∴ the parallelogram $LL' = 4$ the parallelogram PD',

$= 4 AC \cdot BC$,

$= AA' \cdot BB'$.

COR. If PF be drawn at right angles to DCD' meeting CD' in F; then

$$PF \cdot CD' = \text{area of parallelogram } PD',$$
$$= AC \cdot BC.$$

PROP. XXIII.

40. If CP and CD be conjugate diameters, and PF be drawn at right angles to CD meeting CA in G, then

$$PF \cdot PG = BC^2.$$

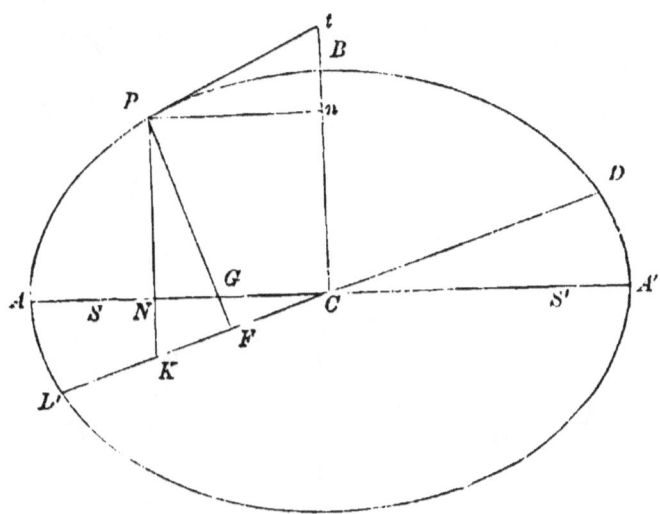

Draw the ordinate PN, and produce it to meet CD' in K.

Also draw Pn at right angles to CB, and let the tangent at P meet CB produced in t.

Now since the angles at N and F are right angles, it is evident that a circle may be described about the quadrilateral figure $NKFG$;

$$\therefore PG \cdot PF = PN \cdot PK, \quad (Euclid, \text{III. 36 } Cor.)$$
$$= Ct \cdot Cn,$$
$$= BC^2. \quad (Prop. \text{ XIV.})$$

Prop. XXIV.

41. If P be any point on the ellipse, and CD be conjugate to CP, then
$$SP \cdot S'P = CD^2.$$

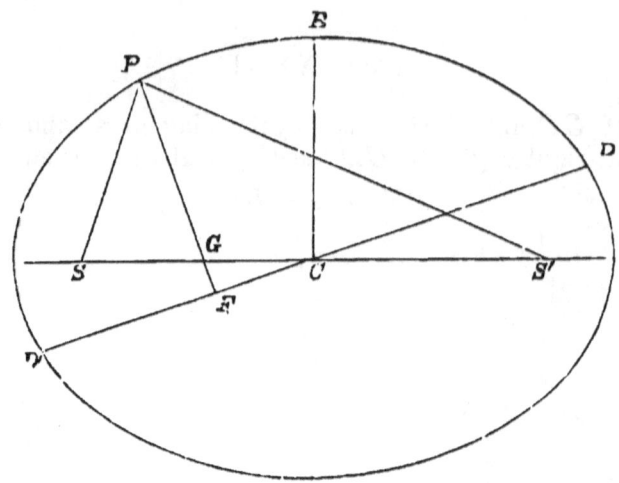

Draw the normal PG and produce it to meet CD' in F; then since CD' is parallel to the tangent at P,

$\therefore PF$ is at right angles to CD',

$\therefore PF \cdot CD = AC \cdot BC$, (Prop. XXII. Cor.)

and $PF \cdot PG = BC^2 = BC \cdot BC$, (Prop XXIII.)

$\therefore CD : PG :: AC : BC$. (1)

Again, $SP : SG :: CA : CS$, (Prop. XI.)

$S'P : S'G :: CA : CS$. (Prop. XI.)

Compounding $SP \cdot S'P : SG \cdot S'G :: CA^2 : CS^2$,

$\therefore SP \cdot S'P : SP \cdot S'P - SG \cdot S'G :: CA^2 : CA^2 - CS^2$.

But $SP \cdot S'P - SG \cdot S'G = PG^2$, (Euclid, VI. Prop. B)

$\therefore SP \cdot S'P : PG^2 :: CA^2 : BC^2$.

But from (1) $CD^2 : PG^2 :: CA^2 : BC$

$\therefore SP \cdot S'P = CD^2$.

This proposition may also be very easily deduced from *Prop.* XV.

Prop. XXV.

42. The area of the ellipse is to the area of the auxiliary circle as BC to AC.

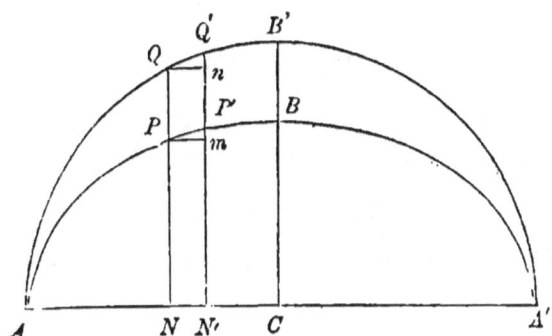

Let PN and $P'N'$ be two ordinates of the ellipse near together.

Produce NP, $N'P'$, to meet the auxiliary circle in Q and Q'.

Draw Pm, Qn, perpendicular to $Q'N'$.

Then

the parallelogram PN' : the parallelogram QN' :: PN : QN,
:: BC : AC.

And the same will be true for all the parallelograms that can be similarly described in the ellipse and auxiliary circle.

Hence the sum of all the parallelograms inscribed in the ellipse is to the sum of all the parallelograms inscribed in the circle as BC to AC.

And this holds however the number of parallelograms be increased.

But when the number of parallelograms is increased, and the breadth of each diminished indefinitely, the sum of the parallelograms inscribed in the ellipse will be equal to the area of the ellipse, and the sum of those inscribed in the circle to the area of the circle. Hence

the area of the ellipse : the area of the circle :: BC : AC.

58   CONIC SECTIONS.

43. DEF. If with a point O on the normal at P as centre, and OP as radius, a circle be described touching the ellipse at P, and cutting it in Q; then when the point Q is made to approach indefinitely near to P, the circle is called the *Circle of Curvature* at the point P.

PROP. XXVI.

If PH be the chord of the circle of curvature at the point P of an ellipse, which passes through the centre; then

$$PH \cdot CP = 2\ CD^2.$$

Let PT be the tangent, and PG the normal at the point P.

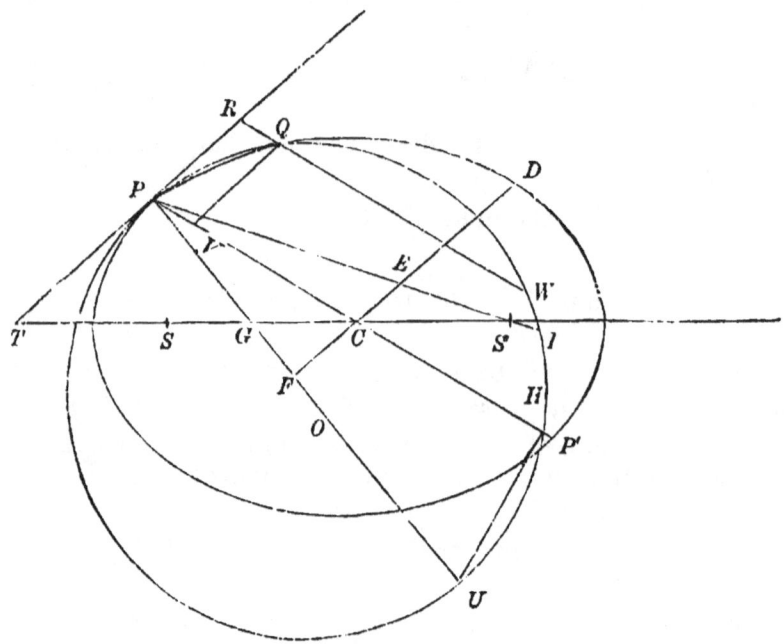

With centre O, and radius OP, describe a circle cutting the ellipse in the point Q.

Draw RQW parallel to CP, meeting the circle in W, and TP produced in R.

Also draw QV parallel to PR, meeting the diameter PP' in V; then since RP touches the circle at P,

$$\therefore RQ \cdot RW = PR^2, \quad (Euclid, \text{III. 36})$$

$$\text{or } PV \cdot RW = QV^2.$$

But $QV^2 : PV \cdot P'V :: CD^2 : CP^2$, (*Prop.* XXI.)

$$\therefore PV \cdot RW : PV \cdot P'V :: CD^2 : CP^2,$$

$$\text{or } RW : P'V :: CD^2 : CP^2.$$

Now, when the circle becomes the circle of curvature at P, the points R and Q move up to, and coincide with P, and the lines RW and PH become equal, while

$P'V$ becomes equal to PP' or $2CP$.

Hence, $PH : 2CP :: CD^2 : CP^2$,

$$\therefore PH \cdot CP : 2CP^2 :: 2CD^2 : 2CP^2,$$

$$\therefore PH \cdot CP = 2CD^2.$$

Prop. XXVII.

If PU be the diameter of the circle of curvature at the point P of the ellipse, and PF be drawn at right angles to CD; then

$$PU \cdot PF = 2CD^2.$$

Since the triangle PHU is similar to the triangle PFC,

$$\therefore PU : PH :: CP : PF,$$

$$\therefore PU \cdot PF = PH \cdot CP,$$

$$= 2CD^2. \quad (\textit{Prop.} \text{ XXVI.})$$

Prop. XXVIII.

If PI be the chord of the circle of curvature through the focus of the ellipse; then

$$PI \cdot AC = 2CD^2.$$

Let PI meet CD in E; then, since the triangles PIU and PEF are similar,

$$\therefore PI : PU :: PF : PE.$$

But $PE = AC$, (*Prop.* XV. *Cor.*)

$$\therefore PI : PU :: PF : AC,$$
$$\therefore PI \cdot AC = PU \cdot PF,$$
$$= 2CD^2. \quad (Prop.\ XXVII.)$$

Prop. XXIX.

44. *If two chords of an ellipse intersect one another, the rectangles contained by their segments are proportional to the squares of the diameters parallel to them.*

Let POP' be any chord drawn through the point O, and let CD be the semi-diameter parallel to it.

Draw the ordinates NP, $N'P'$, MD, and produce them to meet the auxiliary circle in Q, Q', D'; then

since $NP : NQ :: N'P' : N'Q'$, (*Prop.* XIII. *Cor.*)

it is evident that PP' and QQ' will meet the axis produced in the same point T.

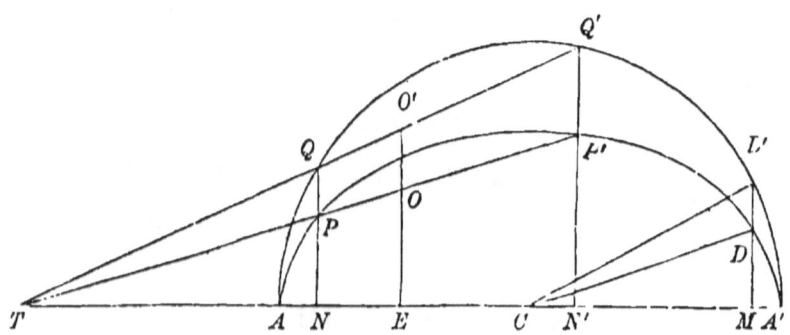

Also since $NP : NQ :: MD : MD'$, (*Prop.* XIII. *Cor.*)

and TPP' is parallel to CD,

$$\therefore TQQ' \text{ is parallel to } CD'.$$

Draw EO parallel to NQ or $N'Q'$, and produce it to meet QQ' in O'; then
$$PO : QO' :: TO : TO',$$
$$\text{and } P'O : Q'O' :: TO : TO',$$
$$\therefore PO \cdot P'O : QO' \cdot Q'O' :: TO^2 : TO'^2,$$
$$:: CD^2 : CD'^2,$$
$$:: CD^2 : AC^2.$$

Alternately, $PO \cdot P'O : CD^2 :: QO' \cdot Q'O' : AC^2$.

Again, if through the point O any other chord pOp' be drawn,
$$\text{since } EO : EO' :: BC : AC,$$
it is manifest that the corresponding chord qq' in the auxiliary circle will pass through the point O'; and if Cd be the semi-diameter parallel to pp' we shall have as before,
$$pO \cdot p'O : Cd^2 :: qO' \cdot q'O' : AC^2.$$
But $QO' \cdot Q'O' = qO' \cdot q'O'$, (*Euclid*, III. 35.)
$$PO \cdot P'O : CD^2 :: pO \cdot p'O : Cd^2,$$
$$PO \cdot P'O : pO \cdot p'O :: CD^2 : Cd^2.$$

The same result may be shown to be true when the point O is without the ellipse.

Prop. XXX.

If QVQ' be any ordinate to the diameter CP, the circle described through the three points P, Q, Q' will intersect the ellipse in a fourth point, which depends only upon the position of P.

Draw the ordinate PN, and produce it to meet the ellipse in P'; then, since, if NT be the subtangent of either P or P',
$$CT \cdot CN = AC^2, (\textit{Prop.} \text{ IX.})$$
therefore the tangents at P and P' will meet the major axis produced in the same point T.

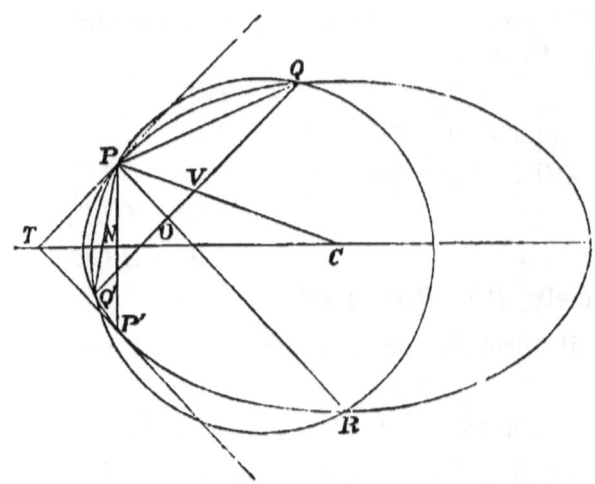

Draw PR parallel to TP', meeting the ellipse in R, and QQ' in O; then if CD and CD' be drawn parallel respectively to TP and TP', meeting the ellipse in D and D',

$$PO \cdot OR : QO \cdot OQ' :: CD'^2 : CD^2. \quad (\textit{Prop. XXIX.})$$

But $CD' = CD$, since $CP' = CP$,

$$\therefore PO \cdot OR = QO \cdot OQ'.$$

Hence, by the converse of *Euclid*, III. *Prop.* 35, the point R is on the circle which passes through P, Q, Q'.

Cor. When the point V is brought indefinitely near to P, QQ' coincides with the tangent to the ellipse at P, and becomes also a tangent to the circle at P since Q and Q' are indefinitely near to each other. The circle therefore becomes the circle of curvature at the point P.

Hence, if PR be drawn parallel to the tangent at P', or be equally inclined to the axis with PT, it will meet the ellipse in the point where the circle of curvature at P intersects the ellipse.

PROBLEMS ON THE ELLIPSE.

1. In what position of P is the angle SPS' greatest?

2. The latus rectum is a third proportional to the axis major and axis minor.

3. Construct on the axis minor as base, a rectangle which shall be to the triangle SLS' in the duplicate ratio of the major axis to the minor axis, L being the extremity of the latus rectum.

4. If a series of ellipses be described having the same major axis; the tangents at the extremities of their latera recta will all meet the minor axis in the same point.

5. Find the locus of the centres of all the ellipses having the same focus, and their major axes of the same length, and touching a given straight line.

6. Given the foci, it is required to describe an ellipse touching a given straight line.

7. If PT be a tangent to an ellipse, meeting the axis in T, and AP, $A'P$, be produced to meet the perpendicular to the major axis through T in Q and Q', then $QT = Q'T$.

8. If the angle SBS' be a right angle, prove that $CA^2 = 2CB^2$.

9. If CP be a semi-diameter, and AQO be drawn parallel to CP meeting the curve in Q, and CB produced in O, then $2CP^2 = AO \cdot AQ$.

10. If AB, CD, which are not parallel, make equal angles with either axis, the lines AC, BD, as also AD, BC, will make equal angles with either axis.

11. PSp is any focal chord. PA and pA are produced to meet the directrix in Q and q. Prove that the angle QSq is a right angle.

12. If a circle be described touching the axis major in one focus, and passing through one extremity of the axis minor; AC will be a mean proportional between the diameter of this circle and BC.

13. If $PQQ'P'$ be a chord of the auxiliary circle, and a circle be described on the minor axis as diameter, cutting the chord in Q and Q', then $PQ \cdot P'Q = CS^2$.

14. If PG be the normal at P, and GL be drawn at right angles to SP, then $PL = \frac{1}{2}$ latus rectum.

15. The sum of the squares of the normals at the extremities of conjugate diameters is constant.

16. If on the normal at P, PQ be taken equal to the semi-conjugate diameter CD, the locus of Q is a circle whose radius is $AC - BC$.

17. Find the locus of the intersection of a pair of tangents at right angles to each other.

18. P is any point on an ellipse. To any point Q on the curve draw AQ, $A'Q$, meeting NP in R and S, and prove that $NR \cdot NS = NP^2$.

19. If PG be a normal, and GL perpendicular to SP, the ratio of GL to PN is constant.

20. If NP produced meet the tangent at the extremity of the latus rectum in Q, then $QN = PS$.

21. In an ellipse the tangent at any point makes a greater angle with the focal distance than with the perpendicular on the directrix.

22. A diameter of an ellipse, parallel to the tangent at any point, meets the focal distances of the point, and from the points of intersection lines are drawn perpendicular to the focal distances. Prove that these lines intersect in the axis minor.

23. The subnormal is a third proportional to CT and BC.

24. If PN be the ordinate of P, prove that $NY : NY'' :: PY : PY'$. (*See fig. Prop.* XV.)

25. If from C lines be drawn parallel and perpendicular to the tangent at P, they inclose a part of one of the focal distances of that point equal to the other.

26. If P be a fixed point on an ellipse, and QQ' an ordinate to CP, the circle $QP'Q'$ will meet the ellipse in a fixed point.

27. P is any point on an ellipse. Draw PP' parallel to the axis major, and through P' draw $P'Q$, $P'Q'$, making equal angles with the major axis. Join QQ'; then QQ' is parallel to the tangent at P.

28. What parallelogram circumscribing an ellipse has the least area?

29. When is the square of the sum of conjugate diameters least?

30. Given the axes of an ellipse, and the position of one focus, and of one point in the curve, give a geometrical construction for finding the centre.

31. If lines drawn through any point of an ellipse to the extremities of any diameter meet the conjugate CD in M and N, then $CM \cdot CN = CD^2$.

32. If CP and CD be conjugate, prove that
$$(SP - AC)^2 + (SD - AC)^2 = SC^2.$$

33. If CP and CD be conjugate, and BP, PD be joined, as also AD, $A'P$, these latter meeting in O, then $BDOP$ is a parallelogram. When is the area greatest?

34. If PSp, QCq be two parallel chords through the focus and centre of an ellipse, prove that
$$SP \cdot Sp : CQ \cdot Cq :: BC^2 : AC^2.$$

35. If the tangent at the vertex A cut any two conjugate diameters in T and t, then $AT \cdot At = BC^2$.

36. If the tangents at three points P, Q, R, intersect in $R_{,}$, $Q_{,}$, $P_{,}$, prove that
$$PR_{,} \cdot P_{,}Q \cdot Q_{,}R = PQ_{,} \cdot R_{,}Q \cdot P_{,}R.$$

F

37. If a circle be described touching SP, $S'P$ produced, and the major axis of the ellipse, find the locus of the centre.

38. If from the extremities of the axes of an ellipse any four parallel lines be drawn, the points in which they cut the curve are the extremities of conjugate diameters.

39. If two equal and similar ellipses have a common centre, the points of intersection are at the extremities of diameters at right angles to one another.

40. If PSQ be a focal chord, and X the foot of the directrix, XP and XQ are equally inclined to the axis.

41. OP, OQ are tangents to an ellipse, and PQ is produced to meet the directrices in R, R', prove that
$$RP \cdot R'P : RQ : R'Q :: OP^2 : OQ^2.$$

42. NPQ is a common ordinate to the ellipse and auxiliary circle. PR, QR are normals at P and Q intersecting in R. The locus of R is a circle whose radius is $AC + BC$.

43. If the conjugate to CP meet SP, $S'P$, or these produced in E, E'; then $SE = S'E'$, and the circles circumscribing SCE, $S'CE'$ are equal.

44. The locus of the middle points of all focal chords in an ellipse is a similar ellipse.

45. The circle described about the triangle SBS' will cut the minor axis in the centre of the circle of curvature at B.

46. The locus of the centre of the circle inscribed in the triangle SPS' is an ellipse.

47. If a circle be described intersecting an ellipse in four points, and chords be drawn through the points of intersection, diameters parallel to the chords will be equal.

48. An ellipse slides between two lines at right angles to each other, find the locus of its centre.

49. If from the focus S perpendiculars be drawn upon the conjugate diameters CP, CD, these perpendiculars produced backward will intersect CD and CP in the directrix.

50. Find the point at which the diameter of curvature is a mean proportional between the major and minor axes.

51. The circle of curvature at a point, where the conjugate diameters are equal, meets the ellipse again at the extremity of the diameter.

52. The locus of the intersection of lines drawn from A, A at right angles to AP, $A'P$ is an ellipse.

53. If a quadrilateral figure be inscribable in two ellipses whose major axes are parallel or perpendicular, any two of its opposite angles will be equal to two right angles.

54. If CN, NP are the abscissa and ordinate of a point P on a circle whose centre is C, and NQ be taken equal to NP, and be inclined to it at a constant angle, the locus of Q is an ellipse.

55. If two ellipses having the same major axes can be inscribed in a parallelogram, the foci will be on the corners of an equiangular parallelogram.

56. Two ellipses, whose major axes are equal, have a common focus. Prove that they intersect in two points only.

57. A circle described about the triangle SPS' cuts the minor axes in R on the opposite side to P. Prove that SR varies as the normal PG.

58. If r and R be the radii of the circles inscribed in and about the triangle SPS, prove that $R \cdot r$ varies as $SP \cdot S'P$.

59. The circle described upon PG as diameter cuts SP, $S'P$ in K and L. Prove that KL is bisected by PG, and is perpendicular to it.

60. If from S' a line be drawn parallel to SP, it will meet SY in the circumference of a circle.

61. T and t are the points where the tangent at P meets the axes. CP is produced to meet in L the circle described about the triangle TCt; prove that PL is half the chord of the circle of curvature at P in the direction of C, and that $CP \cdot CL$ is constant.

62. About the triangle PQR an ellipse is described, having its centre at the point where the lines drawn from P, Q, R, to the middle points of the opposite sides meet. CP, CQ, CR, are produced to meet the ellipse in P', Q', R'. Prove that

the tangents at P', Q', R' form a triangle similar to PQR, and four times as large.

63. Lines from Y and Y' perpendicular to the major axis cut the circles on SP, $S'P$ as diameters in I and J. Prove that IS and JS' when produced, intersect BC in the same point.

64. If from the ends of any diameter chords be drawn to any point in the ellipse, the diameters parallel to these chords will be conjugate.

65. If T be the angle between tangents at the extremities of a focal chord, and O the angle subtended by the chord at the other focus, then
$$2T + O = 2 \text{ right angles.}$$

66. The acute angles which SP, SQ make with the tangents are complementary. Prove that BC^2 is a mean proportional between the areas of the triangles SPS', SQS'. Also, show that the problem is impossible unless $BC < CS$.

67. A series of ellipses have their equal conjugate diameters of the same magnitude. One of these diameters is fixed and common, while the other varies. The tangents drawn from any point in the fixed diameter produced will touch the ellipses in points situated on a circle.

68. If on the longer side of a rectangle as major axis an ellipse be described, passing through the intersection of the diagonals, and lines be drawn from any point of the ellipse exterior to the rectangle to the ends of the remote side, they will divide the major axis into segments, which are in geometric progression.

69. From any point P of an ellipse PQ is drawn at right angles to SP meeting the diameter conjugate to CP in Q. Prove that PQ varies inversely as the perpendicular from P on the major axis.

70. In an ellipse SQ and $S'Q$, drawn at right angles to a pair of conjugate diameters, intersect in Q. Prove that the locus of Q is a concentric ellipse.

CHAPTER III.

THE HYPERBOLA.

45. DEF. The *Hyperbola* is the curve traced out by a point which moves in such a manner that its distance from a given fixed point continually bears the same ratio, *greater than unity*, to its distance from a given fixed line. (See *Introduction*.)

PROP. I.

The focus and directrix of a hyperbola being given, to find any number of points on the curve.

Let S be the focus, and MX the directrix.

Draw SX at right angles to the directrix, and divide SX in the point A, so that SA may be to AX in the given fixed ratio, greater than unity; then

A is a point on the curve.

On SX produced take a point A', such that

$$SA' : A'X :: SA : AX;$$

then A' will also be a point on the curve.

On the directrix take *any* point M; and through S and M draw the line $SYMY''$, meeting AY and $A'Y''$, drawn at right angles to AA', in the points Y and Y'';

On YY'' as diameter describe a circle, and draw PMP parallel to AA', cutting the circle in the points P and P';

P and P' will be points on the hyperbola.

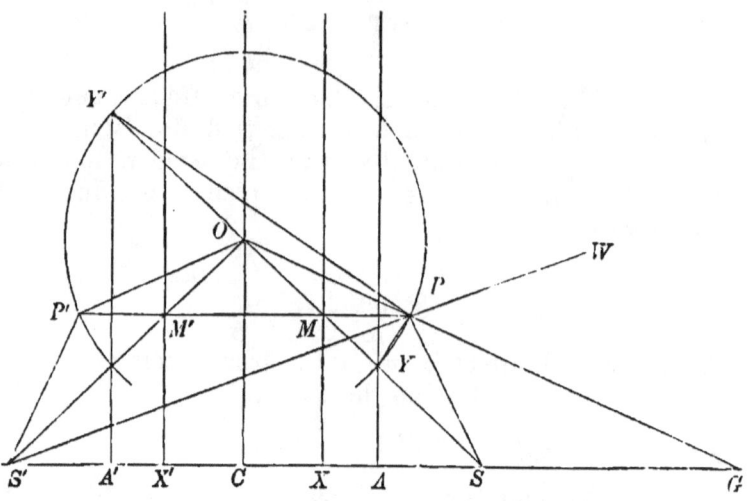

Join PY, PY'', SP; then since
$$SY : YM :: SA : AX, \quad (Euclid,\ VI.\ 2)$$
and $SY' : Y''M :: SA' : A'X, \quad (Euclid,\ VI.\ 2)$
$$\therefore SY : YM :: SY' : Y''M;$$
or, alternately, $SY : SY' :: YM : Y''M$,
and the angle YPY'' in a semicircle is a right angle,
$$\therefore PY \text{ bisects the angle } SPM,{}^*$$
$$\therefore SP : PM :: SY : YM,$$
$$:: SA : AX.$$

So we may show that
$$SP' : P'M :: SY' : Y'M,$$

---

* For, if not, make the angle YPm equal to YPS; then
$$SY : Ym :: SP : Pm. \quad (Euclid,\ VI.\ 1.)$$
and since, if PY bisect SPm, PY' being at right angles to PY, also bisects the angle between MP and SP produced;
$$\therefore SY' : Y'm :: SP : Pm, \quad (Euclid,\ VI.\ A.)$$
Hence $SY : Ym :: SY' : Y'm$,
or $SY : SY' :: Ym : Y'm$,
$\therefore$ the points M and m coincide.

$$:: SA : AX,$$

∴ P and P' are points on the curve.

In the same way, by taking other points on the directrix, we may obtain as many more points on the curve as we please.

Cor. 1. Since, corresponding to every point P on the curve, there is a point P' situated in precisely the same manner with respect to $A'Y'$ as P is with respect to AY, it is clear that if we make $A'S'$ equal to AS, and $A'X'$ equal to AX, and draw $X'M'$ at right angles to AX', the curve could be equally well described with S' as focus and $M'X'$ as directrix.

The hyperbola is therefore symmetrical, not only with respect to the line AA', but also with respect to the line OC drawn through the middle point of YY' at right angles to and bisecting AA'.

Cor. 2. The line OP produced will bisect the angle SPW between SP and $S'P$ produced.

Produce OP and $S'S$ to meet in G. Produce PM to meet $X'M'$ in M', and draw OS' passing through the point M'; then

$$SP : PM :: S'P : PM',$$

or, alternately, $SP : S'P :: PM : PM'$. (1)

Again, $SG : PM :: S'G : PM'$,

or, alternately, $SG : S'G :: PM : PM'$. (2)

∴ from (1) and (2)

$$SP : S'P :: SG : S'G,$$

∴ PG bisects the angle SPW. (*Euclid*, VI. A.)

It will be shown hereafter (*Prop.* IX.) that the normal to the hyperbola at the point P also bisects the angle SPW. Hence the hyperbola and circle have the same tangent at the point P. The hyperbola will consequently touch all the infinite series of circles which can be described in the same manner as the one in the figure, by taking different points on the directrix.

Prop. II.

46. If C be the middle point of AA', then CA is a mean proportional between CS and CX,

or $CS \cdot CX = CA^2$. (See *fig. Prop.* III.)

Since $SA' : A'X :: SA : AX$.

Alternately $SA' : SA :: A'X : AX$,

$\therefore SA' - SA : SA :: A'X - AX : AX$;

or $AA' : SA :: XX' : AX$,

$\therefore AA' : XX' :: SA : AX$,

or $CA : CX :: SA : AX$. (1.)*

Again, $SA' : SA :: A'X : AX$.

$\therefore SA' + SA : SA :: A'X + AX : AX$,

or $SS' : SA :: AA' : AX$.

Alternately, $SS' : AA' :: SA : AX$,

or $CS : CA :: SA : AX$. (2.)

Hence from (1) and (2)

$CA : CX :: CS : CA$,

$\therefore CA^2 = CX \cdot CS$.

or CA is a mean proportional between CS and CX.

Cor. Since the three lines CS, CA, CX, are proportional therefore, by the definition of duplicate ratio and *Euclid*, VI. 20 *Cor.*,

$CS : CX :: CS^2 : CA^2$. (3.)

Prop. III.

47. If P be any point on the hyperbola, and S be the focus nearer to P; then

$S'P - SP = AA'$.

Since $SP : PM :: SA : AX$,

---

* N.B. The results (1), (2), (3), should be remembered, as they will frequently be referred to.

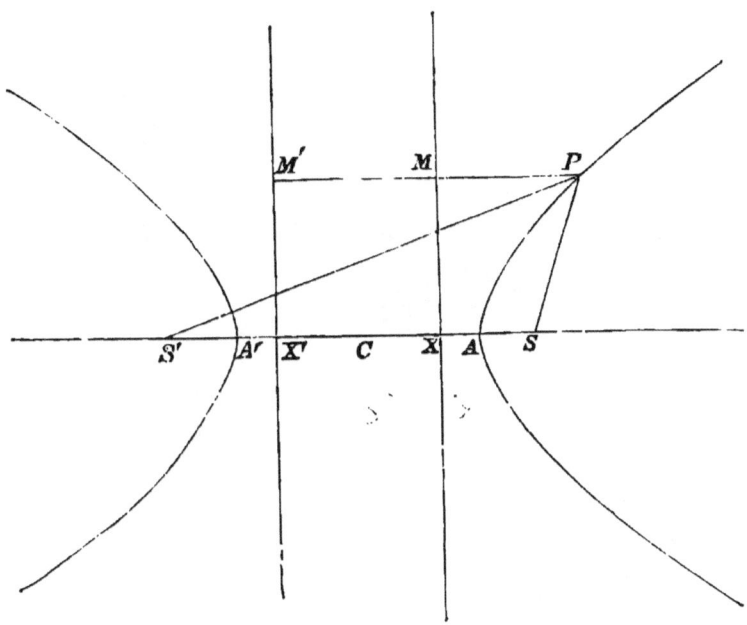

and $SA : AX :: AA' : XX'$,   (*Prop.* II.)

$\therefore SP : PM :: AA' : XX'$.

So $S'P : PM' :: AA' : XX'$,

$\therefore S'P - SP : PM' - PM :: AA' : XX'$.

But $PM' - PM = MM' = XX'$,

$\therefore S'P - SP = AA'$.

Cor. By means of this property the hyperbola may be practically described, and the form of the curve determined.

Let a rigid bar $S'Q$ of any length have one end fastened at the focus S', in such a manner that it is capable of turning freely round S' as a centre in the plane of the paper.

At the other end of the bar let a string be fastened of such a length that when stretched along the bar it shall just reach to within a distance equal to AA' from the end S' of the bar.

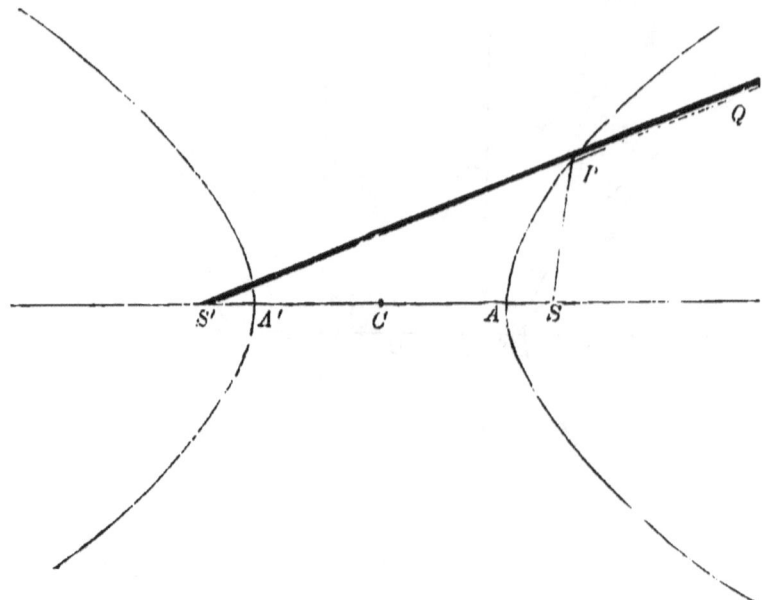

If the loose end of the string be now fastened to the focus S, and the rod being initially placed in the position $S'S$, be made to revolve round S', while the string is kept constantly stretched by means of the point of a pencil at P, in contact with the bar; since $S'P$ and SP are always increasing by the same amount, viz. the length of the portion of the string that removes itself from the bar, between any two positions of P, the difference between $S'P$ and SP will be constantly the same, and the point P will trace out the hyperbola.

Another perfectly similar branch may be described in the same manner by making the bar revolve round S as centre.

In this case $S'P - SP$ will be equal to AA'.

The curve, therefore, consists of two similar branches, which recede indefinitely both from the line AA', and also from the line BCB' drawn bisecting AA' at right angles. (See fig. Prop. IV.)

48. If BC be taken of such a length that
$$BC^2 = CS^2 - CA^2,$$

and CB' be made equal to CB, then AA' and BB' are called respectively the *Transverse* and *Conjugate Axes*.

The line BCB' does not meet the hyperbola, and the reason of its being introduced will be seen further on.

If the conjugate and transverse axes are equal, the hyperbola is said to be *rectangular* or *equilateral*.

The property of the hyperbola, which we have just investigated, viz. that the difference between SP and $S'P$ is constant, is sometimes taken as the definition of the curve. (*See Chapter* II. *Art.* 25.)

Also as in the ellipse, if SL be the semi-latus rectum, it may be proved that
$$SL \cdot AC = BC^2.$$

Prop. IV.

49. *The difference of the distances of any point from the foci of a hyperbola will be greater or less than AA', according as the point is on the concave or convex side of the curve.*

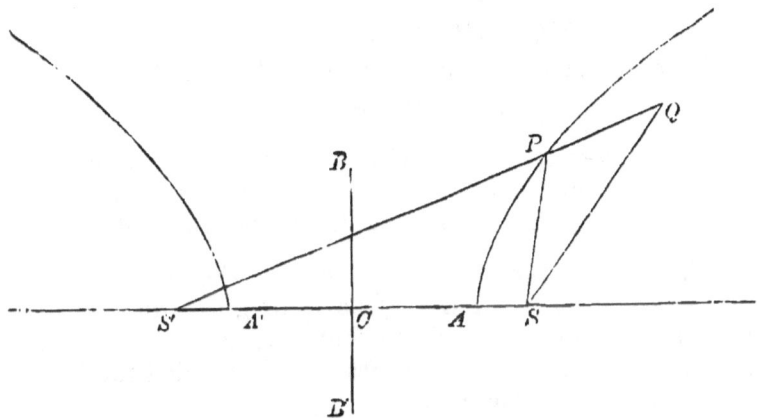

(1.) Let Q be a point on the concave side of the hyperbola. Join SQ, $S'Q$, and let $S'Q$ meet the curve in P; join SP; then
since $S'Q = S'P + PQ$,
and $SQ < SP + PQ$,

76 CONIC SECTIONS.

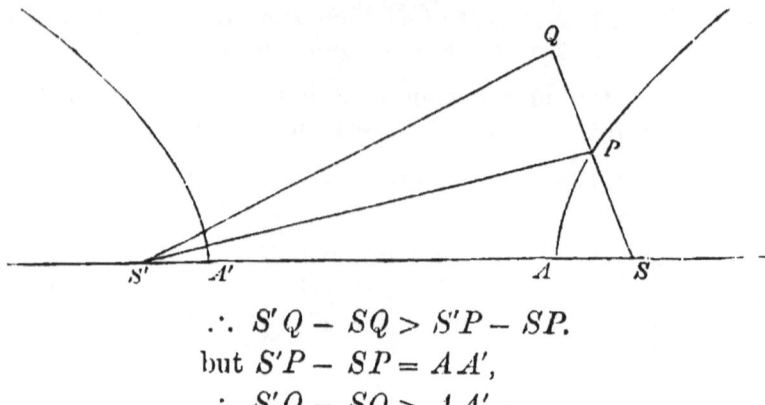

$$\therefore S'Q - SQ > S'P - SP.$$
$$\text{but } S'P - SP = AA',$$
$$\therefore S'Q - SQ > AA'.$$

(2.) Let Q be a point on the convex side of the curve, nearer to S than S'; join SQ, $S'Q$, and let SQ meet the curve in P; join $S'P$; then

$$S'Q < S'P + PQ,$$
$$\text{and } SQ = SP + PQ,$$
$$\therefore S'Q - SQ < S'P - SP,$$
$$\text{but } S'P - SP = AA'.$$
$$\therefore S'Q - SQ < AA',$$

so if Q be nearer to S' than S, we can show that

$$SQ - S'Q < AA';$$

COR. Conversely a point will be on the concave or convex side of the hyperbola, according as the difference of its distances from the foci is greater or less than AA'.

50. DEF. If a point P' be taken on the hyperbola near to P (*see fig. Prop.* V.) and PP' be joined, the line PP' produced, in the limiting position which it assumes when P' is made to approach indefinitely near to P, is called the *Tangent* to the hyperbola at the point P.

PROP. V.

If the tangent to the hyperbola at any point P meet the directrix in the point Z, and if S be the focus corresponding to the directrix on which Z is situated, then SZ will be at right angles to SP.

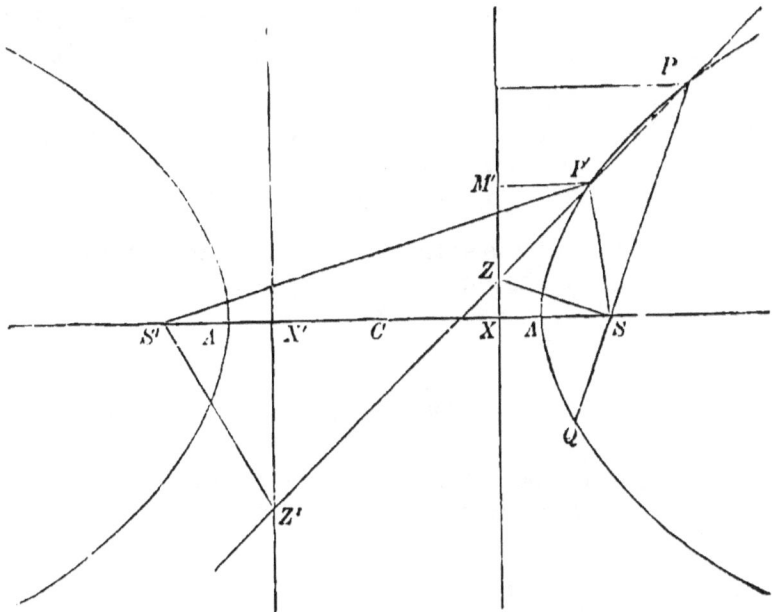

Let P' be a point in the curve near to P.

Draw the chord PP', and produce it to meet the directrix in Z; join SZ.

Draw PM, $P'M'$ at right angles to the directrix, and join SP, SP'.

Produce PS to meet the hyperbola in Q; then since the triangles ZMP, $ZM'P'$ are similar,

$$\therefore ZP : ZP' :: MP : M'P',$$
$$:: SP : SP'.$$

$\therefore SZ$ bisects the angle $P'SQ$. (*Euclid*, VI. A.)

Now, when P' is indefinitely near to P, and PP' becomes the tangent at the point P, the angle PSP' becomes indefinitely small, while the angle QSP' approaches two right angles; and therefore the angles ZSP', being half of the angle $P'SQ$, becomes ultimately a right angle.

Hence, when PZ becomes the tangent at the point P, the angle ZSP is a right angle,

or SZ is perpendicular to SP.

78   CONIC SECTIONS.

Cor. 1. Conversely, if SZ be drawn at right angles to SP, meeting the directrix in Z, and PZ be joined, PZ will be the tangent at P.

Cor. 2. If PZ be produced to meet the other directrix in Z', and $S'Z'$ be joined; then
$$S'Z' \text{ is at right angles to } S'P'.$$

Cor. 3. The tangents at the extremities of the latus rectum, or double ordinate through the focus, meet the axis in the point X.

Prop. VI.

The tangent to the hyperbola at any point P makes equal angles with the focal distances SP and $S'P$.

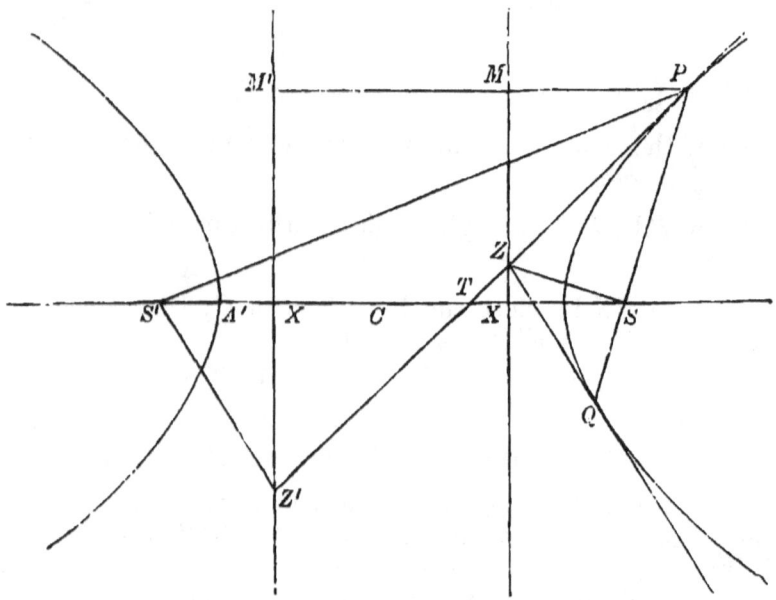

Let the tangent at P meet the directrices in Z and Z'.

Draw PMM' at right angles to the directrices meeting them in M and M' respectively; join SZ, $S'Z'$; then
$$SP : PM :: S'P : PM'.$$

And since the triangles ZMP, $Z'M'P$ are similar,
$$PM : PZ :: PM' : PZ',$$
$$\therefore SP : PZ :: S'P : PZ'. \quad (Ex\ æquali.)$$

Now in the triangles SPZ, $S'PZ'$ because the sides about the angles SPZ, $S'PZ'$ are proportional, and the angles PSZ, $PS'Z'$ are equal, being right angles, and the angles SZP, $S'Z'P$ are each less than a right angle,

$\therefore$ the triangles SPZ, $S'PZ'$ are similar. (*Euclid*, VI. 7).
$$\therefore \text{ the angle } SPZ = S'PZ'.$$

Prop. VII.

The tangents at the extremities of a focal chord intersect in the directrix.

Let PSQ be a focal chord, and let the tangent P meet the directrix in Z. Join SZ; then

the angle ZSP is a right angle, (*Prop.* V.)

And $\therefore$ also the angle ZSQ is a right angle,

$\therefore ZQ$ is the tangent at Q. (*Prop.* V. *Cor.* 1.)

Or the tangents at the extremities of a focal chord intersect in the directrix.

Prop. VIII.

51. If the tangent at P meet the transverse axis in T, and PN be the ordinate of the point P; then
$$CT \cdot CN = CA^2.$$

Draw PMM' at right angles to the directrices meeting them in M and M'. Join SP, $S'P$; then

since PT bisects the angle SPS', (*Prop.* VI.)
$$\therefore S'T : ST :: S'P : SP, (Euclid, \text{VI. 3.})$$
$$:: PM' : PM,$$
$$:: X'N : XN.$$

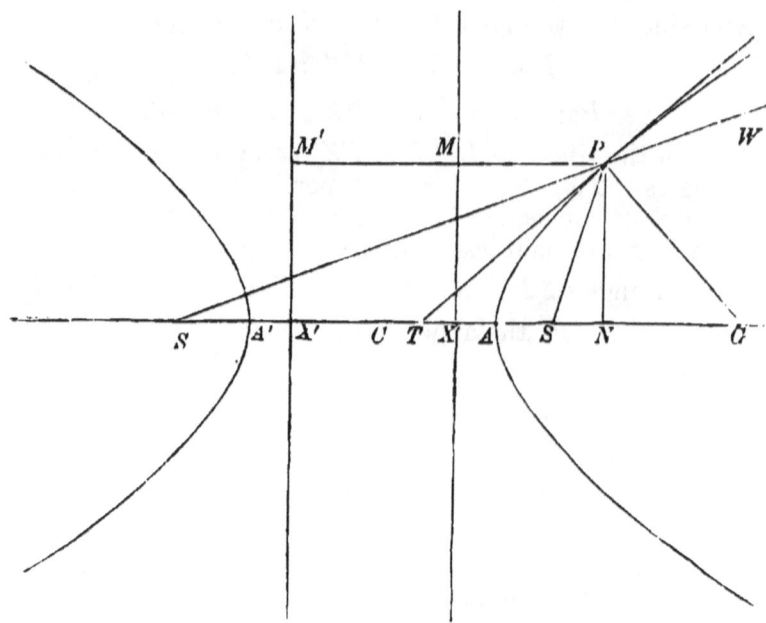

$$\therefore S'T - ST :: S'T + ST :: X'N - XN : X'N + XN$$
$$\text{or } 2\,CT : 2\,CS :: 2\,CX : 2\,CN,$$
$$\text{or } CT : CS :: CX : CN.$$
$$\therefore CT \cdot CN = CS \cdot CX,$$
$$= CA^2. \quad (Prop. \text{ II.})$$

52. DEF. The line PG, drawn at right angles to the tangent PT, is called the *Normal* to the hyperbola at the point P.

PROP. IX.

If the normal to the hyperbola at the point P meet the transverse axis in the point G, and PN be the ordinate of the point P, then

$$NG : NC :: BC^2 : AC^2.$$

Draw PMM' at right angles to the directrices, meeting them in M and M', and produce $S'P$ to W; then since
the angle TPG is a right angle,
$\therefore$ the angle $WPG =$ the complement of the angle $S'PT$, and the angle $SPG =$ the complement of the angle SPT;

but the angle $S'PT$ = the angle SPT,

∴ the angle WPG = the angle SPG,

∴ PG bisects the angle SPW,

∴ $S'G : SG :: S'P : SP$, (*Euclid*, VI. A.)

$:: PM' : PM$,

$:: N'N : NN$,

∴ $S'G + SG : S'G - SG :: N'N + NN : N'N - NN$;

or $2CG : SS' :: 2CN : NN'$.

Alternately, $2CG : 2CN :: SS' : NN$;

or $CG : CN :: CS : CN$,

$:: CS^2 : CA^2$ (*Prop.* II. Cor.)

∴ $CG - CN : CN :: CS^2 - CA^2 : CA^2$;

or $NG : CN :: BC^2 : AC^2$.

Prop. X.

If PN be the ordinate of any point P on the hyperbola, then

$$PN^2 : AN . A'N :: BC^2 : AC^2.$$

For $NG : NC :: BC^2 : AC^2$.

And rectangles of the same altitude are to one another as their bases, (*Euclid*, VI. 1.)

∴ $TN . NG : TN . NC :: BC^2 : AC^2$;

or $PN^2 : TN . NC :: BC^2 : AC^2$.

But $TN . CN = CN^2 - CT . CN$, (*Euclid*, II. 2.)

$= CN^2 - CA^2$, (*Prop.* VIII.)

$= AN . A'N$, (*Euclid*, II. 6)

∴ $PN^2 : AN . A'N :: BC^2 : AC^2$.

Prop. XI.

If the normal at any point P of an hyperbola meet the transverse axis in G; then

$$SG : SP :: CS : CA.$$

Produce $S'P$ to W; then

since PG bisects the angle SPW, (Prop. IX.)
$$SG : S'G :: SP : S'P,$$
$$\therefore SG : S'G - SG :: SP : S'P - SP,$$
but $S'P - SP = AA'$, (Prop. III.)
and $S'G - SG = SS'$,
$$\therefore SG : SS' :: SP : AA',$$
or $SG : SP :: SS' : AA'$,
or $SG : SP :: CS : CA$.

Cor. Hence also,
$$S'G : S'P :: CS : CA.$$

Prop. XII.

53. If from the foci S and S' of an hyperbola SY and $S'Y'$ are drawn at right angles to the tangent at P, then Y and Y' are on the circumference of the circle described on AA' as diameter, and
$$SY . S'Y' = BC^2.$$

Join SP, $S'P$, and produce SY to meet $S'P$ in W; join CY; then

since the angle $SPY =$ the angle WPY, (Prop. VI.)
and the angle $SYP =$ the angle WYP,
and the side PY is common to the triangles SPY, WPY,
$\therefore$ the triangle $SPY = WPY$ in all respects,
$$\therefore SP = PW, \text{ and } SY = WY,$$
$$\therefore S'P - SP = S'W,$$
but $S'P - SP = AA'$, (Prop. III.)
$$\therefore S'W = AA'.$$
Again, $\therefore SC = CS'$, and $SY = WY$,
$$\therefore SC : CS' :: SY : YW,$$
$\therefore CY$ is parallel to $S'W$,
$$\therefore CY : SW :: CS :: SS'.$$

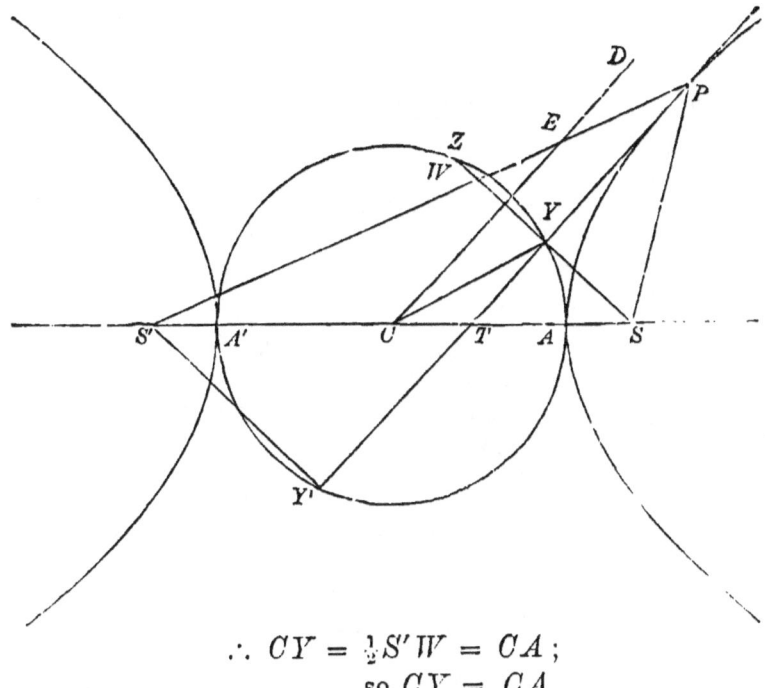

$$\therefore CY = \tfrac{1}{2} S'W = CA;$$
$$\text{so } CY = CA.$$

∴ Y and Y' are points on the circumference of the circle described upon AA' as diameter.

Next, let SY be produced to meet this circle in Z, and join ZY'; then

since the angle ZYY' is a right angle

∴ ZY' passes through the centre C,

∴ the angle SCZ = the angle $S'CY'$,

$$\therefore SZ = S'Y',$$
$$\therefore SY \cdot S'Y' = SY \cdot SZ,$$
$$= SA \cdot SA', \text{ (Euclid, III. 36 Cor.)}$$
$$= CS^2 - CA^2, \text{ (Euclid, II. 6.)}$$
$$= BC^2.$$

Cor. If CD be drawn parallel to the tangent at P meeting $S'P$ in E; then

since $CEPY$ is a parallelogram,

$$\therefore PE = CY = AC.$$

Prop. XIII.

54. To draw a pair of tangents to an hyperbola from an external point O.

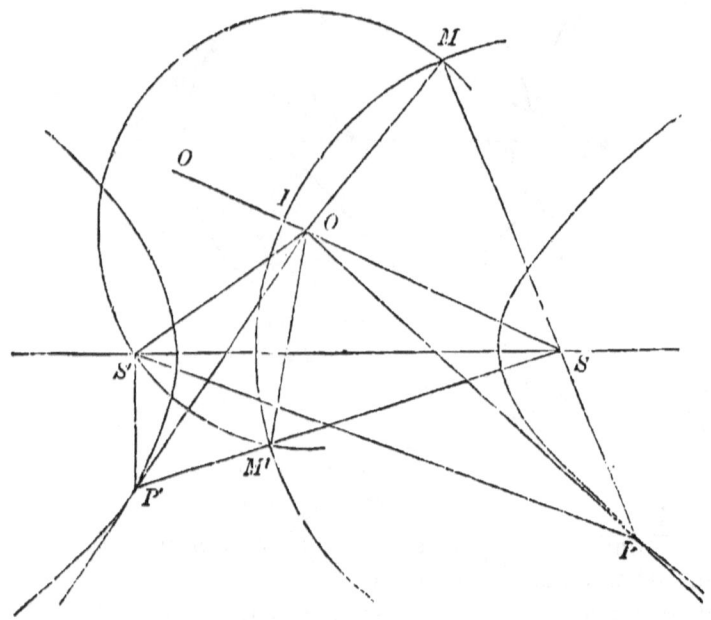

Of the foci S and S', let S' be that which is nearer to O.

With centre S and radius equal to AA' describe a circle.

Join OS, OS'; and let SO or $S'O$ produced meet the circle in the point I.

Now if O be a point inside the circle MIM' it is evident that OS' is greater than OI; and if O be outside the circle,

since $OS - OS' < AA'$ or SI, (*Prop.* IV.)

$\therefore OS - OS' < OS - OI$,

$\therefore OS' > OI$.

With centre O and radius OS' describe another circle cutting the former in the points M and M', which it will always do since OS' is greater than OI.

Join SM, SM', and produce them to meet the hyperbola in the points P and P'.

Join OP, OP'; these will be the tangents required.

Join $S'P$, $S'P'$; then

since $S'P - SP = AA' = SM$,

$\therefore S'P = PM$.

And $\therefore S'P, PO = MP, PO$, each to each,

and $OS'' = OM$,

$\therefore$ the angle OPS' = the angle OPM,

$\therefore OP$ is the tangent at P. (*Prop.* VI.)

So OP is the tangent at P'.

The points of contact P and P' will be upon the same or opposite branches of the hyperbola according as SM and SM' require to be produced in the same or in opposite directions with respect to S, in order to intersect the hyperbola.

Prop. XIV.

If from a point O a pair of tangents, OP, OP' be drawn to an hyperbola, then the angles which OP and OP' subtend at either focus will be equal or supplementary according as the points of contact are in the same or opposite branches of the hyperbola.

Let the points P and P' be on opposite branches of the hyperbola.

Join PS, $S'P$; SP', $S'P'$.

Produce PS to M, making PM equal to PS'. Also from $P'S$ cut off a part $P'M'$ equal to $P'S'$.

Join OM, OM'; OS, OS'.

Then since $OP, PS' = OP, PM$, each to each,

and the angle OPS' = the angle OPM, (*Prop.* VI.)

$\therefore OS' = OM$,

and the angle $OS'P$ = the angle OMP.

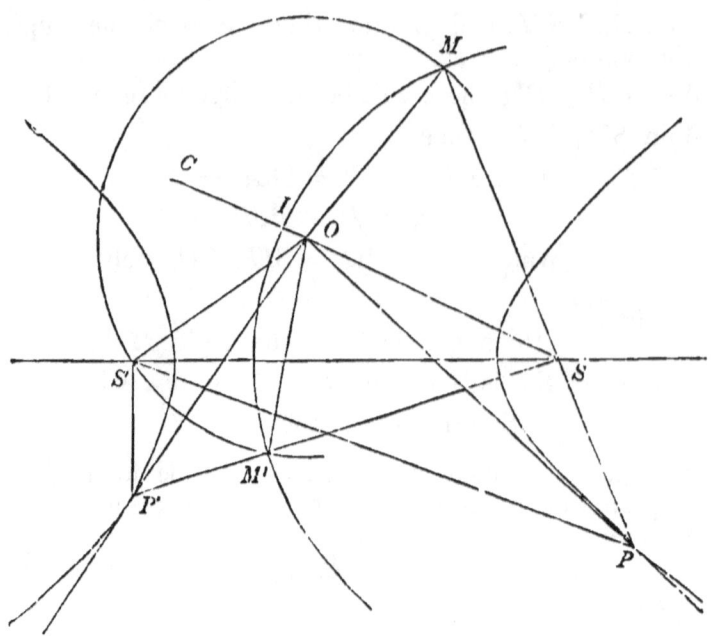

So $OS' = OM'$,
and the angle $OS'P' = $ the angle $OM'P'$,
$\therefore OM = OM'$.
Again, $\because SM = S'P - SP = AA'$,
and $SM' = SP' - S'P' = AA'$,
$\therefore SM = SM'$.
And $\because OS, SM = OS, SM'$, each to each,
and $OM = OM'$,
$\therefore$ the angle $OSM =$ the angle OSM',
and the angle $OMS =$ the angle $OM'S$.
But OSM is the supplement of OSP,
and $OM'S$ is the supplement of $OM'P'$,
$\therefore OSM'$ is the supplement of OSP,
and OMP the supplement of $OM'P'$.
But $OMP = OS'P$,
and $OM'P' = OS'P'$,
$\therefore OS'P$ is the supplement of $OS'P$.

Hence the angles which OP and OP' subtend either at S or S' are supplementary.

In a similar manner if P and P' are on the same branch of the hyperbola, the angles subtended either at S or S' may be shown to be equal.

Prop. XV.

55. If the tangent at any point P of an hyperbola meet the conjugate axis in the point t, and Pn be drawn at right angles to CB; then

$$Cn \cdot Ct = BC^2.$$

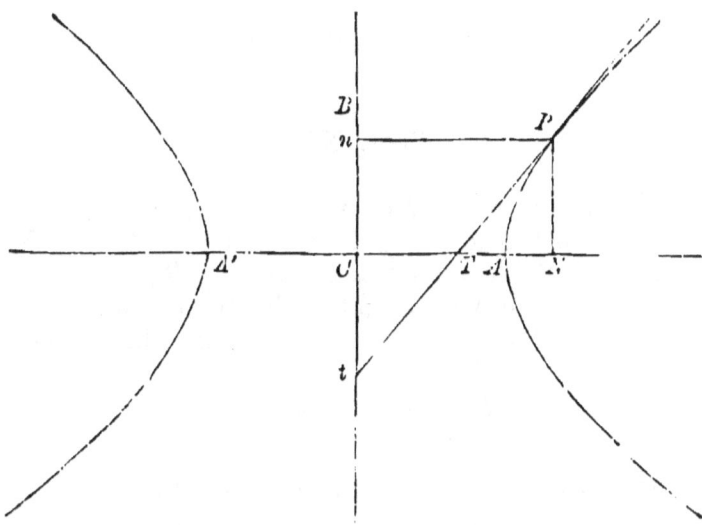

Draw PN at right angles to CA; then
$$Ct : CT :: PN : NT,$$
$$\therefore Ct : PN :: CT : NT,$$
$$\therefore Ct \cdot Cn : PN^2 :: CT \cdot CN : CN \cdot NT;$$
or $Ct \cdot Cn : CT \cdot CN :: PN^2 : CN \cdot NT,$
$$:: BC^2 : AC^2. \quad (Prop.\ X.)$$
But $CT \cdot CN = AC^2,$
$$\therefore Ct \cdot Cn = BC^2.$$

56. The proofs that we have given up to this point of the properties of the hyperbola are closely analogous to the corresponding propositions in the ellipse. The remaining properties of the hyperbola are more conveniently investigated by means of its relation to certain lines, which we shall presently define, called *Asymptotes*, in the same manner as many of the properties of the ellipse were deduced from those of the auxiliary circle.

DEF. The hyperbola described (*see fig. Prop.* XIV.) with C as centre, and BB' as transverse axis, and AA' as conjugate axis, is called the *Conjugate Hyperbola*. Its foci, which will be on the line BCB', will evidently be at the same distance from C as those of the original hyperbola, since
$$CS^2 = CA^2 + CB^2.$$

PROP. XVI.

If through any point R or either of the diagonals of the rectangle formed by drawing tangents to the hyperbola and its conjugate at the vertices, A, A', B, B', two ordinates RPN, RDM, be drawn at right angles to AA' and BB', and meeting either the hyperbola or its conjugate in the points P and D; then
$$RN^2 - PN^2 = BC^2,$$
$$\text{and } RM^2 - DM^2 = AC^2.$$

Let R be a point on the diagonal $O'CO$; then
$$RN^2 : CN^2 :: AO^2 : AC^2,$$
$$:: BC^2 : AC^2,$$
and $PN^2 : CN^2 - CA^2 :: BC^2 : AC^2$; (*Prop.* X.)
$$\therefore RN^2 - PN^2 : CA^2 :: BC^2 : AC^2;$$
$$\therefore RN^2 - PN^2 = BC^2.$$
Again, $RM^2 : CM^2 :: AC^2 : BC^2,$
and $DM^2 : CM^2 - CB^2 :: AC^2 : BC^2$; (*Prop.* X.)
$$\therefore RM^2 - DM^2 : BC^2 :: AC^2 : BC^2;$$
$$\therefore RM^2 - DM^2 = AC^2.$$

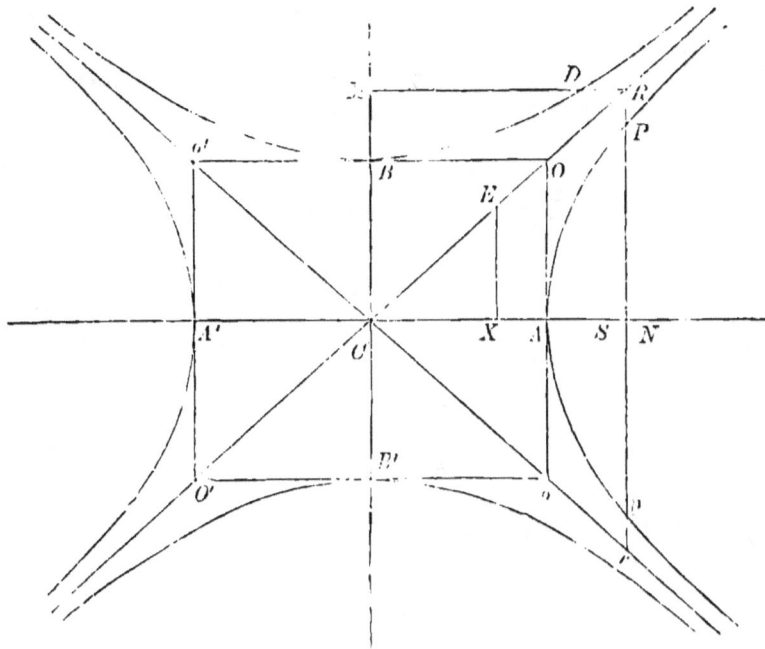

In exactly the same manner, if NR had been produced to meet the conjugate hyperbola in P, and MR had been produced to meet the original hyperbola in D, we should have had,

$$PN^2 - RN^2 = BC^2,$$
$$\text{and } DM^2 - RM^2 = AC^2.$$

Cor. If RP be produced to meet the hyperbola in p, and the other asymptote in r; then

$$RN^2 - PN^2 = RP \cdot Pr; \text{ (Euclid, II. 5.)}$$
$$\therefore RP \cdot Pr = BC^2.$$

Hence as RPN is further removed from A, and the line Pr consequently increases, since the rectangle contained by RP and Pr remains constant, RP must diminish, and by taking R sufficiently far from C, RP may be made less than any assignable magnitude. The line CR, therefore, continually approaches nearer and nearer to the hyperbola, though it never actually reaches it.

On account of this property, CR is called an *Asymptote* to the hyperbola.

So also if P be the point where NR produced meets the conjugate hyperbola, we shall have

$$RP \cdot Pr = BC^2;$$

and therefore CR is also the asymptote to the conjugate hyperbola.

In the same manner it may be shown that the other diameter oCo' of the rectangle OO' is an asymptote to both hyperbolas.

Prop. XVII.

57. If E be the point where the asymptote meets the directrix; then

$$CE = AC.$$

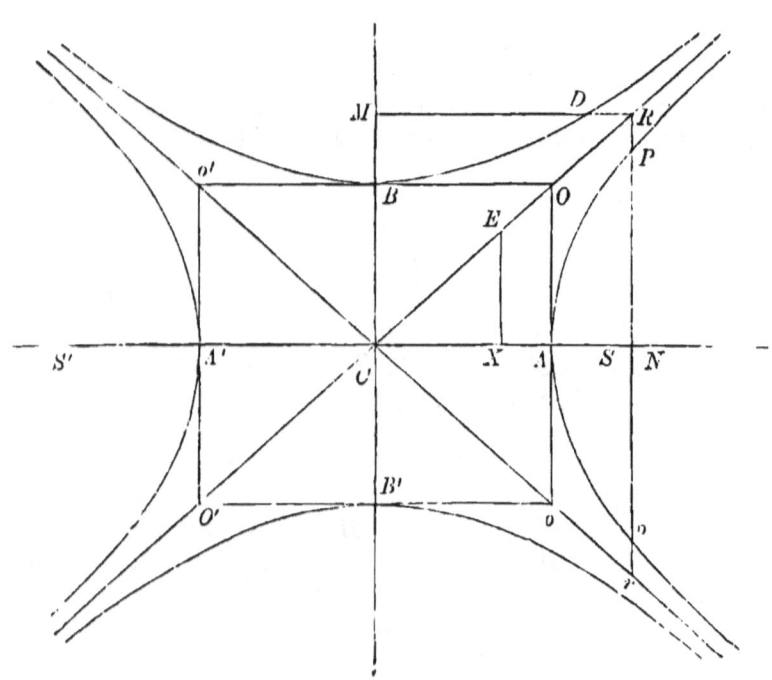

For by similar triangles,
$$CE : CO :: CX : CA,$$
$$:: CA : CS. \quad (Prop.\ II.)$$
But $CO^2 = CA^2 + CB^2 = CS^2;$
$$\therefore CO = CS;$$
$$\therefore CE = AC.$$

Cor. If SE be joined, since
$$CE^2 = CA^2 = CS \cdot CX,$$
$\therefore$ the angle CES is a right angle. (*Euclid*, VI. 8, *Cor.*)

Prop. XVIII.

If from any point R in one of the asymptotes to an hyperbola ordinates RPN, RDM be drawn to the hyperbola and its conjugate respectively, and PD be joined, PD will be parallel to the other asymptote.

For $RN^2 : RM^2 :: BC^2 : AC^2;$
and $RN^2 - PN^2 : RM^2 - DM^2 :: BC^2 : AC^2,$ (*Prop.* XVI.)
$$\therefore PN^2 : DM^2 :: BC^2 : AC^2;$$
$$:: RN^2 : RM^2,$$
$$\therefore PN : DM :: RN : RM;$$
$\therefore PD$ is parallel to MN. (*Euclid*, VI. 2.)
Also $CN : CM :: AC : BC,$
$\therefore MN$ is parallel to $AB;$
and $OA : Ao :: OB : Bo',$
$\therefore AB$ is parallel to $oo'.$
Hence PD is parallel to $oo'.$

Cor. So also if R and D be the points where NR and MR produced meet respectively the conjugate and the original hyperbola, PD will be still parallel to $oo'.$

Prop. XIX.

58. If through any two points Q and Q' of an hyperbola a line $RQQ'R'$ be drawn in any direction meeting the asymptotes in R and R'; then will

$$RQ = R'Q.$$

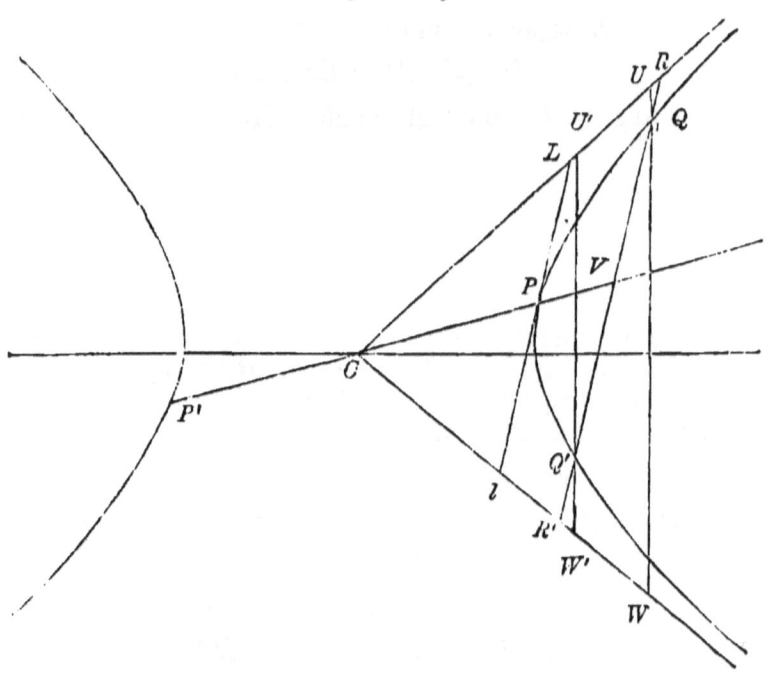

Through Q and Q' draw the ordinates UQW, $U'Q'W'$; meeting the asymptotes in U, W, U', W'; then by similar triangles,

$$QR : QU :: Q'R : Q'U',$$
and $QR' : QW :: Q'R' : Q'W'$;

∴ compounding

$$QR \cdot QR' : QU \cdot QW :: Q'R \cdot Q'R' : Q'U' \cdot Q'W'.$$

But $QU \cdot QW = BC^2 = Q'U' \cdot Q'W$, (*Prop.* XVI. *Cor.*)

$$\therefore QR \cdot QR' = Q'R \cdot Q'R';$$

but $QR \cdot QR' = QR \cdot QQ' + QR \cdot Q'R'$,
and $Q'R \cdot Q'R' = Q'R' \cdot QQ' + QR \cdot Q'R'$;
$\therefore QR \cdot QQ' = Q'R' \cdot QQ'$.
$\therefore QR = Q'R'$.

Cor. 1. If $RQQ'R'$ move parallel to itself until the points Q and Q' coincide, the line RQR' will ultimately assume the position LPl, and will become a tangent to the hyperbola at P.

Hence, since RQ is always equal to $R'Q'$,
$$LP = Pl,$$
or the tangent LPl is bisected at the point of contact P.

Cor. 2. If CP be produced to meet RR' in V, then since
$RV : VR' :: LP : Pl$,
$\therefore RV = VR'$;
and $RQ = Q'R'$,
$\therefore QV = Q'V$.

Hence, if a series of parallel chords be drawn in an hyperbola, their middle points will all be in the line drawn through the centre and the point where the tangent parallel to the chords meets the hyperbola.

Def. A line PCP' drawn through the centre, and meeting the hyperbola in P and P', is called a *Diameter*.

A diameter consequently bisects all chords drawn parallel to the tangents at its extremities.

Prop. XX.

59. If through any point Q of an hyperbola a line RQr be drawn in any direction meeting the asymptotes in R and r, and LPl be the tangent drawn parallel to RQr; then
$$RQ \cdot Qr = PL^2.$$

Through P and Q draw the ordinates EPe, UQW, meeting the asymptote in E, e, U, W; then by similar triangles,

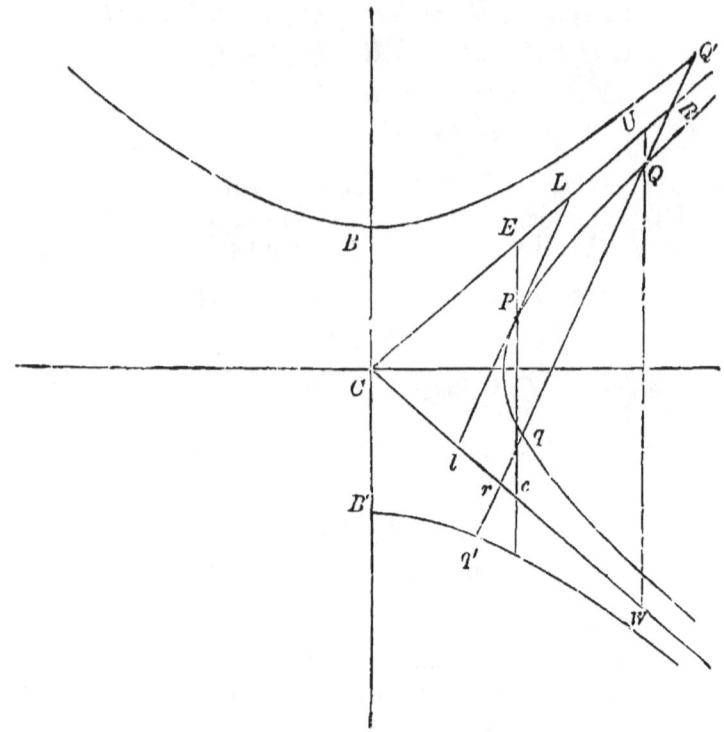

$$QR : QU :: PL : PE,$$
$$Qr : QW :: Pl : Pc;$$
$$\therefore QR \cdot Qr : QU \cdot QW :: PL \cdot Pl : PE \cdot Pc;$$
but $QU \cdot QW = BC^2 = PE \cdot Pc$, (*Prop.* XVI. *Cor.*)
$$\therefore QR \cdot Qr = PL \cdot Pl,$$
$$= PL^2. \quad (Prop. \text{ XIX. } Cor. \text{ 1.})$$

Cor. If Qq be produced to meet the conjugate hyperbola in Q', q', we may show that
$$Q'R \cdot Q'r = PL^2,$$
and also, as in Proposition XIX., that
$$Q'R = q'r,$$
$$\therefore QQ' = qq'.$$

Hence if a line be drawn in any direction meeting both the hyperbolas, the portions intercepted between the hyperbola and its conjugate will be equal.

Prop. XXI.

60. If from any point P of an hyperbola, PH and PK be drawn parallel to the asymptotes, meeting them in H and K respectively; then $4 \cdot PH : PK = CS^2$.

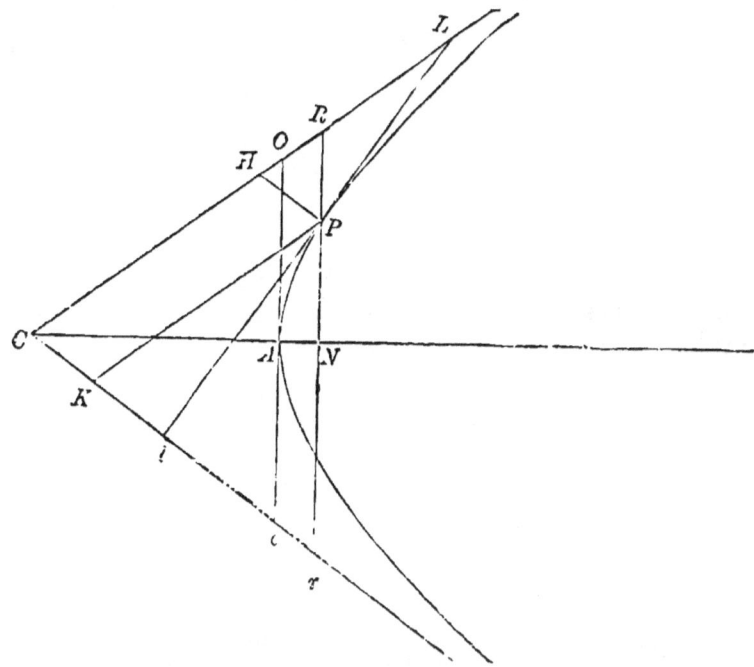

Draw the ordinate $RPNr$ meeting the asymptotes in R and r; then by similar triangles,

$$PH : PR :: Co : Oo,$$
$$\text{and } PK : Pr :: CO : Oo,$$
$$\therefore PH \cdot PK : PR \cdot Pr :: CO^2 : Oo^2,$$
$$:: CS^2 : 4BC^2.$$
But $PR \cdot Pr = BC^2$,
$$\therefore 4 \cdot PH \cdot PK = CS^2.$$

Prop. XXII.

If the tangent at any point P of an hyperbola meet the asymptotes in L and l; then the area of the triangle LCl is equal to the rectangle contained by AC and BC.

Draw PH and PK parallel to the asymptotes meeting them in H and K; then

since $CL : CH :: Ll : Pl$,

and $Ll = 2Pl$, (*Prop.* XIX. *Cor.* 1.)

$\therefore CL = 2CH = 2PK$;

so $Cl = 2CK = 2PH$,

$\therefore CL . Cl = 4PH . PK = CS^2$, (*Prop.* XXI.)

$= CO . Co$,

$\therefore CL : CO :: Co : Cl$,

$\therefore$ the triangles LCl, OCo have the angle at C common and the sides about those angles reciprocally proportional·

$\therefore$ the triangle $LCl =$ the triangle OCo,

$= AC . AO$.

$= AC . BC$.

Prop. XXIII.

61. If from any point R in the asymptote of an hyperbola two ordinates RPN and RDM be drawn to the hyperbola and its conjugate respectively, then the tangents at P and D will be parallel respectively to CD and CP.

Join PD, meeting CR in H; then

since PD is parallel to oo', (*Prop.* XVIII.)

the tangents at P and D will each meet CR produced in the same point L. (*Prop.* XXII.)

Produce LP and LD to meet the other asymptotes in l and l'; then

since $CL . Cl = CS^2 = CL . Cl'$, (*Prop.* XXII.)

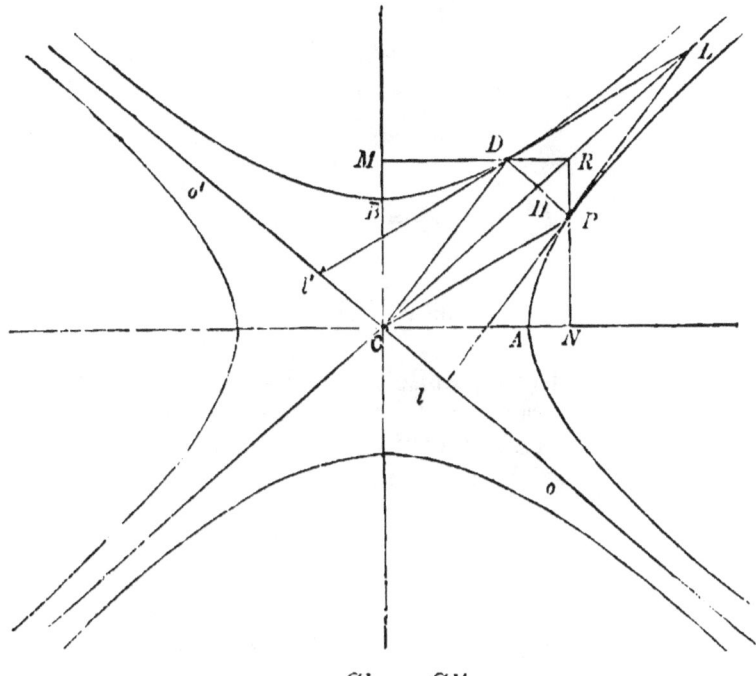

$$\therefore Cl = Cl',$$
$$\therefore lC : Cl' :: lP : PL,$$
$$\therefore CP \text{ is parallel to the tangent at } D.$$
$$\text{Also } l'D : DL :: l'C : Cl,$$
$$\therefore CD \text{ is parallel to the tangent at } P.$$

The lines CP and CD are called *Conjugate Diameters*, since each of these lines is parallel to the tangent at the extremity of the other.

Prop. XXIV.

If CP and CD be semi-conjugate diameters in the hyperbola; then
$$CP^2 \backsim CD^2 = CA^2 \backsim CB^2.$$

Draw the ordinates NPR, MDR meeting the asymptote in the point R (*Prop.* XXIII.); then

H

$$CR^2 - CP^2 = NR^2 - NP^2,$$
$$= BC^2, \quad (Prop. \text{ XVI.})$$
$$\therefore CR^2 = CP^2 + BC^2;$$
$$\text{so } CR^2 = CD^2 + AC^2,$$
$$\therefore CP^2 + BC^2 = CD^2 + AC^2;$$
$$\text{or } CP^2 \backsim CD^2 = AC^2 \backsim BC^2.$$

Prop. XXV.

62. The area of any parallelogram formed by drawing tangents to the hyperbola and its conjugate at the extremities P, P', D, D' of a pair of conjugate diameters is equal to the rectangle contained by the axes.

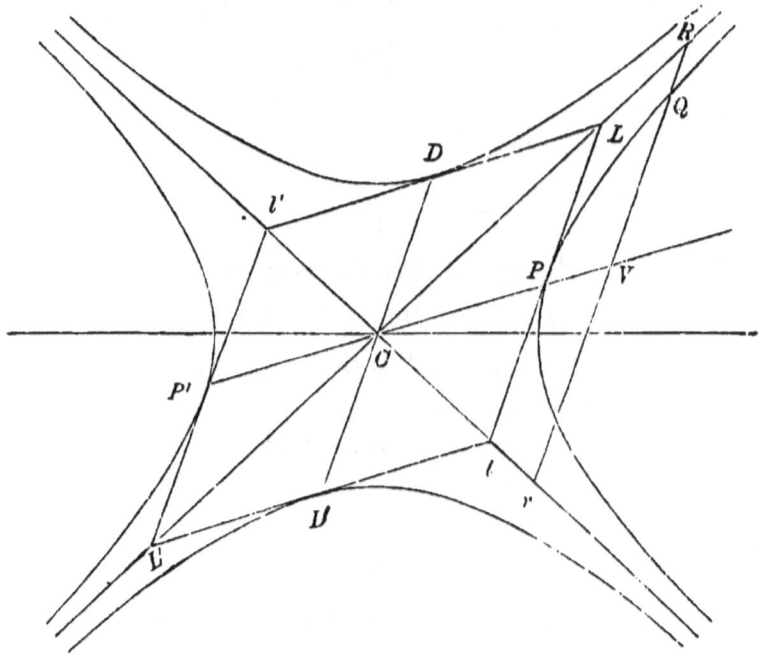

Let Ll, $L'l'$ be the parallelogram formed by drawing tangents at the extremities P, P', D, D', of any pair of conjugate diameters. The points L, L', l, l', will (*Prop. XXIII.*) be on the asymptotes.

Now the parallelogram $LL' = 4$ parallelogram CL,
$$= 4 \text{ triangle } LCl,$$
$$= 4 AC . BC, \quad (Prop. \text{ XXII.})$$
$$= AA' . BB'.$$

Cor. If PF be drawn perpendicular to CD, then
$$PF . CD = AC . BC.$$

Also, if the normal PG meet the transverse axis in G, as in the ellipse
$$PF . PG = BC^2.$$

63. Def. The line QV drawn from any point Q of the hyperbola parallel to the tangent at any point P, and meeting CP produced in V, is called an *Ordinate* to the diameter CP.

Prop. XXVI.

If QV be an ordinate to the diameter $P'CP$, and CD be conjugate to CP; then
$$QV^2 : PV . P'V :: CD^2 : CP^2.$$

Produce VQ to meet the asymptotes in R and r; and let the tangent at P meet the asymptotes in L and l; then
$$RV^2 : PL^2 :: CV^2 : CP^2,$$
$$\therefore RV^2 - PL^2 : PL^2 :: CV^2 - CP^2 : CP^2.$$
But $RQ . Qr = PL^2$, (Prop. XX.)
$$\therefore RV^2 - QV^2 = PL^2,$$
or $RV^2 - PL^2 = QV^2$.
And $CV^2 - CP^2 = PV . P'V$, (*Euclid.* II. 6.)
$$\therefore QV^2 : PL^2 :: PV . P'V : CP^2.$$
Alternately, $QV^2 : PV . P'V :: PL^2 : CP^2$.
But since PD is a parallelogram, (Prop. XXIII.)
$$\therefore PL = CD.$$
Hence $QV^2 : PV . P'V :: CD^2 : CP^2$.

Cor. If VQ be produced to meet the conjugate hyperbola in Q', then
since $Q'R . Q'r = PL^2$, (*Prop.* XX. *Cor.*)
$$\therefore Q'V^2 - RV^2 = PL^2.$$
Hence $Q'V^2 : CV^2 + CP^2 :: CD^2 : CP^2$.

Prop. XXVII.

64. If QV be an ordinate to the diameter PV, and the tangent at Q meet CP in the point T; then
$$CV \cdot CT = CP^2.$$

Draw the tangent LPl meeting the asymptotes in the points L, l; also let the tangent at Q meet the asymptotes in R, r.

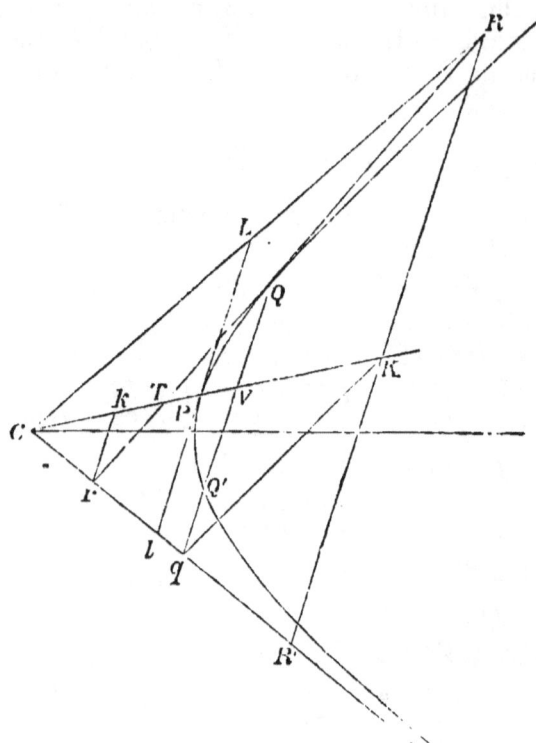

Draw RK, rk, parallel to QV meeting CP in K, k.

Now since the triangles RCr, LCl are equal, (*Prop.* XXII.) and have the angle at C common,

$$\therefore CR : CL :: Cl : Cr. \quad (\textit{Euclid}, \text{VI. 15.})$$

But $CR : CL :: CK : CP$,

and $Cl : Cr :: CP : Ck$,

$\therefore CK : CP :: CP : Ck$,

$\therefore CK \cdot Ck = CP^2$.

Again, produce RK and QV to meet the asymptote Cl in R' and q; then

Since Rr is bisected in Q, (*Prop.* XIX. *Cor.* 1.)

$\therefore R'r$ is bisected in q,

and $RK = R'K$, (*Prop.* XIX. *Cor.* 2.)

$\therefore Kq$ is parallel to Rr,

$\therefore CT : CK :: Cr : Cq$,

$:: Ck : CV$,

$\therefore CV \cdot CT = CK \cdot Ck$

$= CP^2$.

Cor. 1. Conversely, if QV be an ordinate to PV,

and $CV \cdot CT = CP^2$,

then QT is the tangent at Q.

Cor. 2. Hence also, if RR' meet the curve in U and U',

and kU, kU' be drawn,

since $CK \cdot Ck = CP^2$,

$\therefore kU$ and kU' are tangents to the hyperbola at U and U'.

Prop. XXVIII.

65. *If two chords of a hyperbola intersect one another, the rectangles contained by their segments are proportional to the squares of the diameters parallel to them.*

Let QOq be any chord drawn through the point O, and let CD be drawn parallel to it, meeting the conjugate hyperbola in D.

Produce Qq to meet the asymptotes in R and r; and draw the diameter CPV, bisecting both Qq and Rr in V. (*Prop.* XIX. *Cor.* 2.)

Also draw the tangent LPl parallel to Qq, meeting the asymptotes in L and l.

102   CONIC SECTIONS.

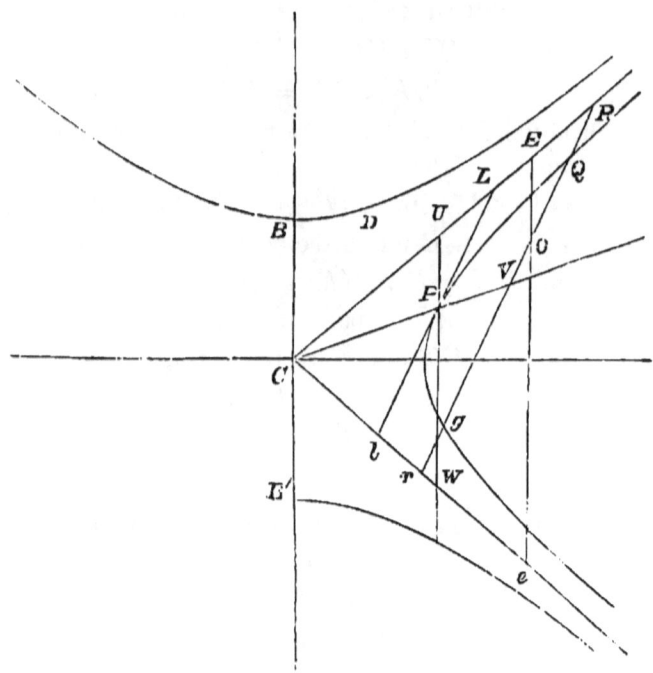

Now since Qq is divided equally in V and unequally in O.

$\therefore QO \cdot Oq = QV^2 - OV^2$; (*Euclid*, II. 5.)

so also $RO \cdot Or = RV^2 - OV^2$; (*Euclid*, II. 5.)

$\therefore RO \cdot Or - QO \cdot Oq = RV^2 - QV^2$,

$\qquad\qquad\qquad = RQ \cdot Qr$ (*Euclid*, II. 5.)

$\qquad\qquad\qquad = PL^2$, (*Prop.* XX.)

$\therefore QO \cdot Oq = RO \cdot Or - PL^2$.

Again, through O and P draw EOe, UPW, at right angles to the axis meeting the asymptotes in E, e, U, W; then

$\qquad\qquad RO : OE :: PL : PU$,

$\qquad\text{and } rO : Oe :: Pl : PW$,

$\therefore RO \cdot rO : OE \cdot Oe :: PL^2 : PU \cdot PW$;

but $PU \cdot PW = BC^2$, (*Prop.* XVI.)

and $PL^2 = CD^2$, (*Prop.* XXIII.)

$\therefore RO \cdot rO : OE \cdot Oe :: CD^2 : BC^2$,

or $RO \cdot rO : CD^2 :: OE \cdot Oe : BC^2$.

$\therefore RO \cdot rO - PL^2 : CD^2 :: OE \cdot Oe - BC^2 : BC^2$,

or $QO \cdot Oq : CD^2 :: OE \cdot Oe - BC^2 : BC^2$.

In the same manner if through O another chord $Q'\,Oq'$ be drawn, and CD' be drawn parallel to it, meeting the conjugate hyperbola in D', we shall have

$$Q'O \cdot Oq' : CD'^2 :: OE \cdot Oe - BC^2 : BC^2.$$

Hence $QO \cdot Oq : Q'O \cdot Oq' :: CD^2 : CD'^2$.

Cor. The same result may be shown to be true when the point O is outside the hyperbola. Moreover, it is not necessary that the chords should be drawn meeting *one* branch only of the hyperbola or the *same* branch. The proportion still holds good when one or both of the chords meet both branches of the hyperbola, or when the chords are drawn in different branches.

66. Def. If with a point O on the normal at P as centre, and OP as radius, a circle be described touching the hyperbola at P, and cutting it in Q; then when the point Q is made to approach indefinitely near to P, the circle is called the *Circle of Curvature* at the point P.

Prop. XXIX.

If PH be the chord of the circle of curvature at the point P of a hyperbola, which passes through the centre; then

$$PH \cdot CP = 2CD^2.$$

Let PT be the tangent, and PG the normal at the point P.

With centre O, and radius OP, describe a circle cutting the hyperbola in the point Q.

Draw RQW parallel to CP, meeting the circle in W, and TP produced in R.

Also, draw QV parallel to PR, meeting the diameter PP' in V; then, since RP touches the circle at P,

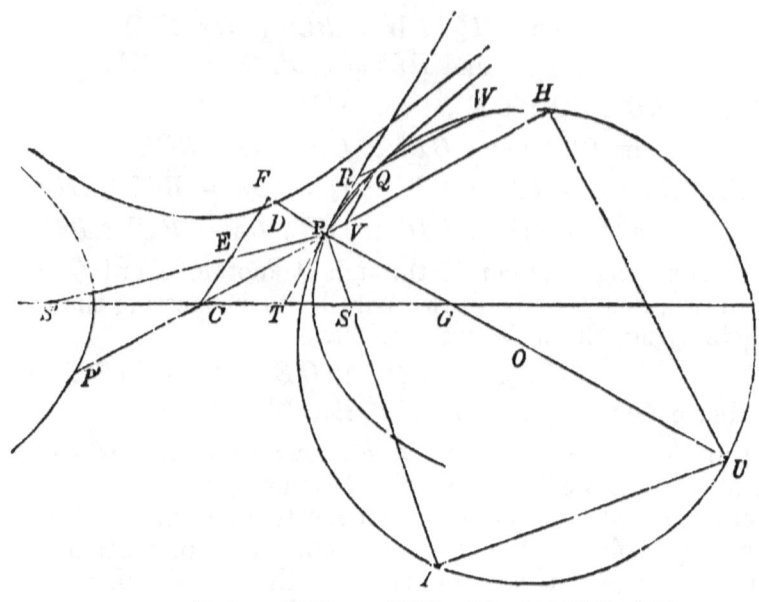

$$\therefore RQ \cdot RW = PR^2, \text{ (Euclid, III. 36.)}$$
$$\text{or } PV \cdot RW = QV^2.$$
But $QV^2 : PV \cdot P'V :: CD^2 : CP^2$, (*Prop.* XXVI.)
$$\therefore PV \cdot RW : PV \cdot P'V :: CD^2 : CP^2,$$
$$\text{or } RW : P'V :: CD^2 : CP^2.$$

Now, when the circle becomes the circle of curvature at P, the points R and Q move up to, and coincide with P, and the lines RW and PH become equal, while

$P'V$ becomes equal to PP', or $2CP$.

Hence, $PH : 2CP :: CD^2 : CP^2$,
$$\therefore PH \cdot CP : 2CP^2 :: 2CD^2 : 2CP^2,$$
$$\therefore PH \cdot CP = 2CD^2.$$

Prop. XXX.

If PU be the diameter of the circle of curvature at the point P of the hyperbola, and PF be drawn at right angles to CD; then

$$PU \cdot PF = 2CD^2.$$

Since the triangle PIU is similar to the triangle PFC,

$$\therefore PU : PH :: CP : PF,$$
$$\therefore PU . PF = PH : CP,$$
$$= 2CD^2. \quad (Prop.\ \text{XXIX}.)$$

Prop. XXXI.

If PI be the chord of the circle of curvature through the focus of the hyperbola; then

$$PI . AC = 2CD^2.$$

Let $S'P$ meet CD in E; then, since the triangles PIU and PEF are similar,

$$\therefore PI : PU :: PF : PE.$$
$$\text{But } PE = AC,\ (Prop.\ \text{XII. Cor.})$$
$$\therefore PI : PU :: PF : AC,$$
$$\therefore PI . AC = PU . PF,$$
$$= 2CD^2. \quad (Prop.\ \text{XXX}.)$$

The point where the circle of curvature intersects the hyperbola may be determined as in the case of the ellipse.

Prop. XXXII.

67. If P be any point on the hyperbola, and CD be conjugate to CP; then

$$SP . S'P = CD^2.$$

Draw PII' parallel to the asymptote CE meeting the directrices in I and I', and CB' in U.

Let the ordinates, NP, MD meet the asymptote in R, and draw PW perpendicular to the directrix; then by similar triangles,

$$PI : PW :: CE : CN,$$
$$:: CA : CN. \quad (Prop.\ \text{XVII}.)$$

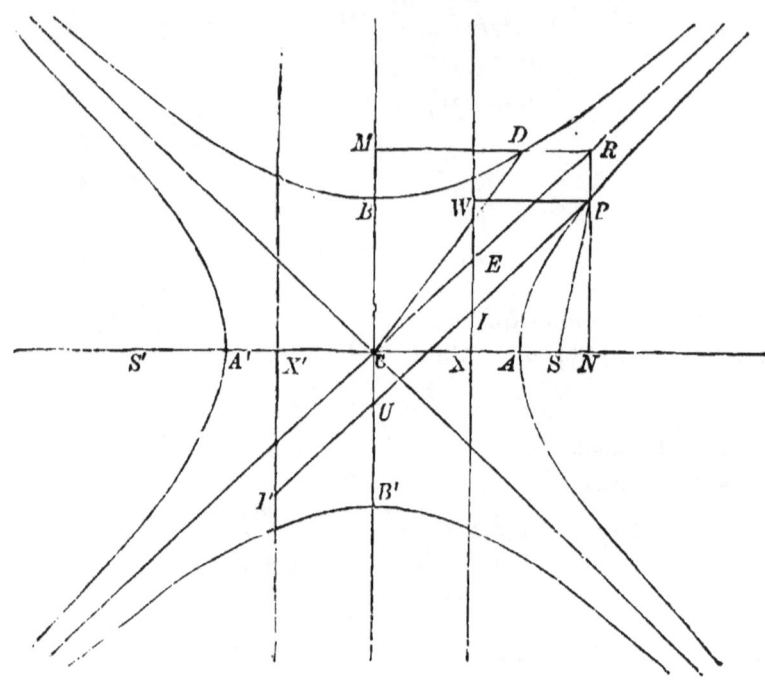

But $SP : PW :: SA : AX$,
$$:: CA : CX.$$
$$\therefore SP = PI;$$
so $S'P = PI'$,
$$\therefore SP \cdot S'P = PI \cdot PI',$$
$$= UP^2 - UI^2,$$
$$= CR^2 - CE^2,$$
$$= CR^2 - CA^2. \text{ (Prop. XVII.)}$$
But $CR^2 - CD^2 = RM^2 - DM^2$,
$$= CA^2, \text{ (Prop. XVI.)}$$
$$\therefore CR^2 - CA^2 = CD^2,$$
Hence $SP \cdot S'P = CD^2$.

PROBLEMS ON THE HYPERBOLA.

1. The locus of the centre of a circle touching two given circles is an hyperbola or ellipse.

2. If on the portion of any tangent intercepted between the tangents at the vertices a circle be described, it will pass through the foci.

3. In an hyperbola the tangents at the vertices will meet the asymptotes in the circumference of the circle described on SS' as diameter.

4. If from a point P in an hyperbola PII' be drawn parallel to the transverse axis meeting the asymptotes in I and I', then $PI \cdot PI' = AC^2$.

5. If a circle be inscribed in the triangle SPS', the locus of its centre is the tangent at the vertex.

6. If PN be the ordinate of the point P, and NQ a tangent to the circle described on the transverse axis as diameter, and PM be drawn parallel to QC meeting the axis in M, then $MN = BC$.

7. If PN be the ordinate of a point P, and NQ be drawn parallel to AP to meet CP in Q, then AQ is parallel to the tangent at P.

8. If an hyperbola and an ellipse have the same foci, they cut one another at right angles.

9. If the tangent at P intersect the tangents at the vertices in R, r, and the tangent at P' intersect them in R', r', then $AR \cdot Ar = AR' \cdot Ar'$.

10. If any two tangents be drawn to an hyperbola, the lines joining the points where they intersect the asymptotes will be parallel.

11. The perpendicular drawn from the focus to the asymptotes of an hyperbola is equal to the semi-conjugate axis.

12. If the asymptotes meet the tangent at the vertex in O, and the directrix in E; then AE is parallel to SO.

13. In a rectangular hyperbola conjugate diameters are equal to one another.

14. In a rectangular hyperbola the normal PG is equal to CP.

15. The lines drawn from any point in a rectangular hyperbola to the extremities of a diameter make equal angles with the asymptotes.

16. Prove that the asymptotes to an hyperbola bisect the lines joining the extremities of conjugate diameters.

17. A line drawn through one of the vertices of an hyperbola and terminated by two lines drawn through the other vertex parallel to the asymptotes will be bisected at the other point where it cuts the hyperbola.

18. P is any point on an hyperbola, and P' a point on the conjugate hyperbola. If CP and CP' be conjugate, prove that
$$S'P' - SP = AC - BC,$$
S and S' being the interior foci.

19. If CP and CD be conjugate, and through C a line be drawn parallel to either focal distance of P, the perpendicular from D upon this line is equal to BC.

20. Given a pair of conjugate diameters, find the principal axes.

21. If Q be a point on the conjugate axis of a rectangular hyperbola, and QP be drawn parallel to the transverse axis meeting the curve in P, then
$$PQ = AQ.$$

22. In a rectangular hyperbola the focal chords drawn parallel to conjugate diameters are equal.

23. If in an equilateral hyperbola CY be drawn at right angles to the tangent at P, and AY be joined, the triangles PCA, CAY are similar.

24. The radius of the circle which touches an hyperbola and its asymptotes, is equal to that part of the latus rectum produced which is intercepted between the curve and the asymptotes.

25. If QQ' be any chord of an hyperbola, and CP the diameter corresponding to it, and QH, PK, $Q'H'$ be drawn parallel to one asymptote meeting the other in H, K and H', then $CH \cdot CH' = CK^2$.

26. If the chord $RPP'R'$ intersect the hyperbola in the points P, P', and the asymptotes in R, R'; and PK be drawn parallel to CR', and $P'K'$ to CR; then $RK = P'K'$, and $R'K' = PK$.

27. If AA' be any diameter of a circle, and PNQ an ordinate to it, then the locus of the intersections of AP, $A'Q$ is a rectangular hyperbola.

28. If two concentric rectangular hyperbolas be described, the axes of one being the asymptotes of the other, they will intersect at right angles.

29. If any chord AP through the vertex be divided in Q, so that $AQ : QP :: AC^2 : BC^2$, and QN be drawn to the foot of the ordinate PN, prove that a straight line drawn at right angles to QN from Q cuts the transverse axis in the same ratio.

30. Prove that the curve which trisects the arcs of all segments of a circle described upon a given base is an hyperbola.

31. If SVs, TVt be two tangents cutting one asymptote in S, T, and the other in s, t, prove that
$$VS : Vs :: Vt : VT.$$

32. If from the exterior focus of an hyperbola a circle be described with radius equal to BC, and tangents be drawn to it from any point in the hyperbola, the line joining the points of contact will touch the circle described on the transverse axis as diameter.

33. Circles are drawn touching the straight line AB in a fixed point C; and from the fixed points A, B tangents are drawn to these circles. The locus of their intersection is an ellipse or hyperbola. Distinguish between the two cases.

34. PP' is a double ordinate in an ellipse. AP, $A'P'$ are produced to meet in Q. Prove that the locus of Q is an hyperbola with the same axes as the ellipse.

35. If the tangent at P intersect the asymptotes in L and l, and PG be the normal at P, then the angle LGl is a right angle.

36. If an ellipse, a parabola, and a hyperbola, have a common tangent, and the same curvature at the vertex, the ellipse will be entirely within the parabola, and the parabola entirely within the hyperbola.

37. The chord $RPP'R'$ of an hyperbola intersects the asymptotes in R and R'. From the point R a tangent RQ is drawn meeting the hyperbola in Q. If PH, QK, $P'H'$ be drawn parallel to one asymptote meeting the other in the points H, K, H'; then $PH + P'H' = 2QK$.

38. If through P,P' on an hyperbola lines be drawn parallel to the asymptotes forming a parallelogram, of which PP is one diagonal, the other diagonal will pass through the centre.

39. If P be the middle point of a line EF which moves so as to cut off a constant area from the corner of a rectangle, its locus is an equilateral hyperbola.

40. PM, PN are drawn parallel to the asymptotes CN, CM, and an ellipse is constructed having CN, CM for semi-conjugate diameters. If CP cut the ellipse in Q, the tangents at Q and P to the ellipse and hyperbola are parallel.

41. If a circle be described through any point P of a given hyperbola and the extremities A, A' of the transverse axis, and NP be produced to meet the circle in Q; prove that Q traces out an hyperbola whose conjugate axis is a third proportional to the conjugate and transverse axes of the original hyperbola.

42. If lines be drawn from any point of a rectangular hyperbola to the extremities of a diameter, the difference between the angles which they make with the diameter will be equal to the angle which this diameter makes with its conjugate.

43. If between a rectangular hyperbola and its asymptotes any number of concentric elliptic quadrants be inscribed, the rectangle contained by their axes will be constant.

44. In the rectangular hyperbola if CP be produced to Q so that $PQ = CP$, and QO be drawn at right angles to CQ to intersect the normal in O, O is the centre of curvature at P.

45. With two conjugate diameters of an ellipse as asymptotes a pair of conjugate hyperbolas are constructed; prove that if one hyperbola touch the ellipse the other will do so likewise; prove also that the diameters drawn through the points of contact are conjugate to each other.

46. If a pair of conjugate diameters of an ellipse when produced be asymptotes to an hyperbola, the points of the hyperbola at which a tangent to the hyperbola will also be a tangent to the ellipse, lie in an ellipse similar to the given one.

47. In the rectangular hyperbola the radius of curvature at P is to the radius of curvature at P' in the triplicate ratio of CP to CP'.

48. OP, OQ are tangents to an ellipse at P and Q, and asymptotes to an hyperbola. Show that a pair of their common chords is parallel to PQ. One of these chords being RS, prove that if PR touches the hyperbola at P, QS touches it at S; also if PS, QR meet in U, OU bisects PQ.

49. The base of the triangle ABC remains fixed, while the vertex C moves in an equilateral hyperbola passing through A and B. If P, Q be the points in which AC, BC meet the circle described on AB as diameter, the intersection of AQ, BP is on the other branch of the hyperbola.

CHAPTER IV.

THE SECTIONS OF THE CONE.

68. DEF. If two indefinite straight lines IOI', DOD', intersect one another at a point O, and one of them IOI' remain fixed while the other DOD' revolves round it in such a manner that its inclination to IOI' is the same in all positions, the surface generated by DOD' will be a *Right Cone*.

The line IOI' is called the *Axis*, and the point O the *Vertex* of the *Cone*.

It now remains for us to show (*see Introduction*) that the curve formed by the intersection of this surface with a plane is *in general* one of the three curves whose properties we have been investigating, and to consider under what circumstances it will be the Parabola, Ellipse, or Hyperbola.

If the cutting plane pass through the vertex of the cone as KOK', and intersect the cone again *at all*, it will *in general* cut it in two straight lines as OK, OK', which will represent two positions of the generating line.

The inclination of these lines to each other will depend upon the inclination of the cutting plane to the axis of the cone, and will be greatest when this plane passes through the axis, in which case it will be double the constant angle between the axis and the generating line.

If the cutting plane pass through a generating line dod' and be perpendicular to the plane containing this line and the axis, it will simply touch the cone along this line.

CONIC SECTIONS. 113

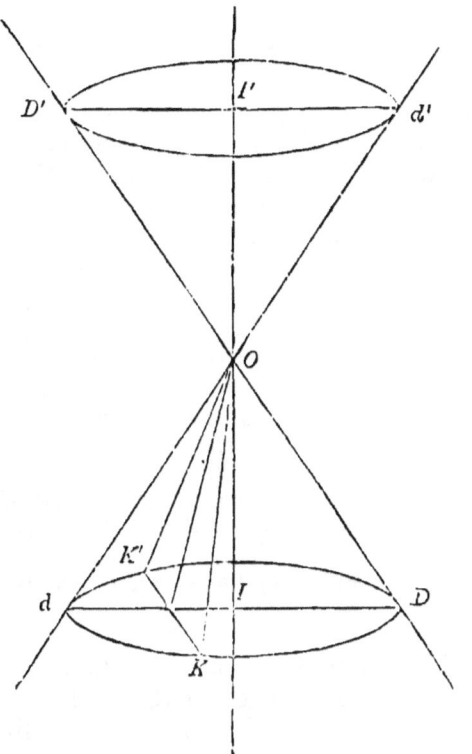

Should the cutting plane not pass through the vertex, and be at right angles to the axis of the cone, the section will evidently be a circle.

In any other case the section will, as we proceed to show, be a Parabola, Ellipse, or Hyperbola.

Whatever be the position of the cutting plane with respect to the cone, we can always suppose a plane drawn through the axis of the cone at right angles to it; and it will be convenient to have this latter plane represented by the plane of the paper as DOd. The cutting plane will therefore always be taken at right angles to the plane DOd of the paper.

Prop. I.

69. The curve formed by the intersection of the surface of a right cone with a plane (which neither passes through its vertex nor is at right angles to its axis) will be a *Parabola, Ellipse,* or *Hyperbola,* according as the inclination of the cutting plane to the axis of the cone is *equal to, greater,* or *less than* the constant angle which the generating line forms with the axis.

Let the plane of the paper represent the plane drawn through the axis IOI' of the cone at right angles to the cutting plane; and let it intersect the surface of the cone in the two generating lines OD, Od.

Let the cutting plane intersect the surface of the cone in the curve PA, and the plane of the paper in the line ANH.

The curve will evidently be symmetrical with respect to this line.

On AH take any point N, and through N draw a plane perpendicular to the axis meeting the surface of the cone in the circle RPr, and the cutting plane in the line PN, which will be at right angles to the plane of the paper and to AN.

Let a sphere be inscribed in the cone touching the cone in the circle EQe and the cutting plane in the point S, and let the plane EQe intersect the cutting plane in the line XM, which will be at right angles to the plane of the paper, and therefore parallel to PN.

Draw PM perpendicular to XM, and join PS, PO, and let PO meet the circle EQe in the point Q.*

Then since PS and PQ are both tangents to the sphere,

$$\therefore PS = PQ.$$
$$\text{But } PQ = RE,$$
$$\therefore PS = RE.$$

---

* N.B. In the figure, to avoid confusion, that part of the section which lies above the plane of the paper is alone represented.

CONIC SECTIONS. 115

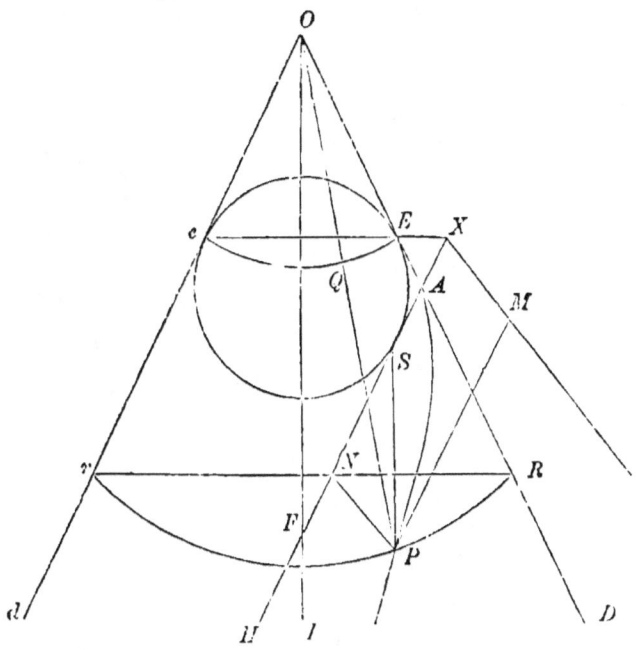

But $RE : XN :: AE : AX$, (*Euclid*, VI. 2.)
and $AE = AS$,
∴ $RE : XN :: AS : AX$,
∴ $SP : PM :: AS : AX$,

∴ the curve PA is either a *Parabola*, *Ellipse*, or *Hyperbola*, whose focus is S and directrix XM.

Again, let AH meet the axis OI in F.

Then the angle AFO will be the inclination of the cutting plane to the axis.

(1) Let the angle AFO = the angle FOd;
then AH is parallel to Od,
∴ the angle AXE = the angle OeE = the angle AEX,
∴ $AE = AX$,
∴ $AS = AX$,
∴ the curve AP will be a *Parabola*.

116 CONIC SECTIONS.

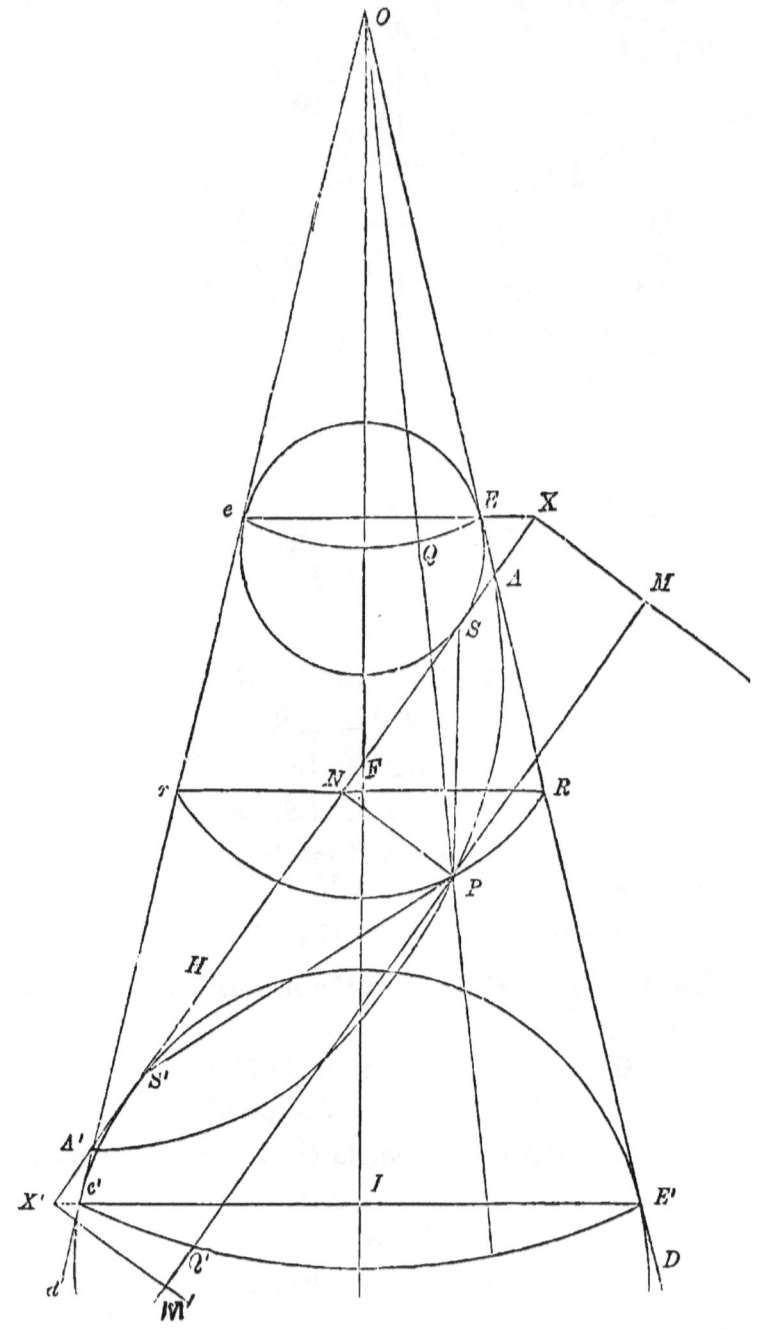

(2) Let the angle AFO be $> FOd$; then

the complement FXE is $<$ the complement OEe,

∴ the angle AXE is $< AEX$,

∴ AE is $< AX$,

or AS is $< AX$,

∴ the curve AP is an *Ellipse*.

Since the angles HFO, FOd are together less than HFO, OFA, *i.e.* than two right angles, the lines AH and Oe may be produced to meet in A'.

If another sphere be described touching the cone in the circle $E'Q'e'$ and the cutting plane in the point S'; and the line $X'M'$ denote the intersection of the plane $E'Q'e'$ with the cutting plane, and PM' be drawn at right angles to this line, it can easily be shown that

$$S'P : PM' :: S'A' : A'X'.$$

Hence S' and $X'M$ represent respectively the other focus and directrix of the ellipse.

Also if BC be the semi-axis minor, and through the centre C a line UCU' be drawn parallel to Ee meeting OD, Od in U and U', then it is evident that

$$BC^2 = CU \cdot CU'.$$

(3) Let the angle AFO be $< FOd$; then

the angle AXE is $>$ the angle AEX,

∴ AE is $> AX$,

∴ AS is $> AX$,

∴ the curve PA is an *Hyperbola*.

Since the angles AFO, FOd' are less than the two FOd, FOd', *i.e.* than two right angles, the lines FA and dO may be produced to meet in A'.

In this case the cutting plane will intersect the other half of the cone, and if any point P' be taken on this part of the curve, and $P'M$ be drawn at right angles to XM, it can be shown as before that

$$SP' : P'M :: S'A :: AX.$$

CONIC SECTIONS.

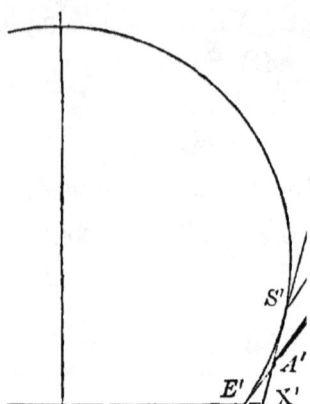

The intersection of the cutting plane therefore with this portion of the cone will be the other branch of the hyperbola.

Also if another sphere be described touching the upper portion of the cone in $E'Q'e'$, and the cutting plane in S', and the line $X'M'$ denote the intersection of the plane $E'Q'e'$ with the cutting plane, and $P'M'$ be drawn at right angles to this line, it can be easily shown that

$$S'M' : P'M' :: S'A' : A'X'.$$

Hence, S' and $X'M'$ will represent respectively the other focus and directrix of the hyperbola.

Cor. 1. In this last case, *i.e.* when the section is an hyperbola, if a plane OKL be drawn through the vertex of the cone parallel to the cutting plane, meeting the plane of the paper in the straight line OL, and the surface of the cone in the generating line OK; then

$$OL : OK :: OL : OR,$$
$$:: AN : AR,$$
$$:: AX : AE, \text{ (Euclid, VI. 2.)}$$
$$:: AX : AS,$$
$$:: CA : CS, \text{ (Chap. III. Prop. II.)}$$

where C is the middle point of AA' and therefore the centre of the hyperbola.

$\therefore KOL$ is half the angle between the asymptotes. (*Chap.* III. *Prop.* XVI.)

Again, if BC be the conjugate semi-axis, and $CU'U$ be drawn parallel to Rr meeting OD', Od' in U and U', then

$$\text{since } CU : AC :: RL : OL,$$
$$\text{and } CU' : A'C :: rL : OL,$$
$$\therefore CU . CU' : AC^2 :: RL . rL : OL^2,$$
$$:: KL^2 : OL^2;$$
$$\text{but } BC^2 : AC^2 :: KL^2 : OL^2,$$
$$\therefore BC^2 = CU . CU'.$$

Cor. 2. If the cutting plane is parallel to the axis OL and OI coincide.

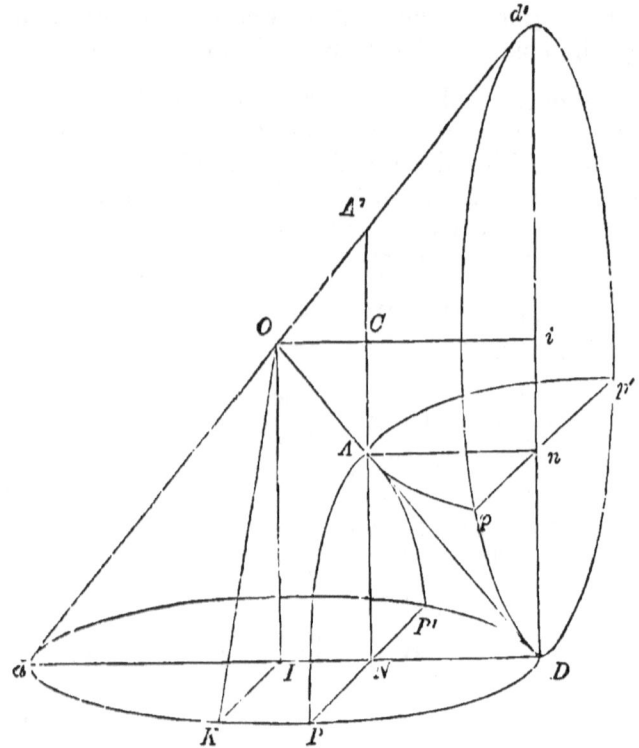

In this case half the angle between the asymptotes of the hyperbolic section is equal to the constant angle DOI, and we can at once see that OC is the semi-conjugate axis.

This affords a convenient method of obtaining a pair of conjugate hyperbolas.

Draw Oi at right angles to OI in the plane of the paper, and let another cone be formed by supposing OD to revolve round Oi in such a manner that the angle DOi is the same in all positions, and equal to the complement of DOI.

Then if through any point A on the common generating line OD we draw two planes at right angles to the plane of the paper, and parallel respectively to OI and Oi, they will cut the cones in two hyperbolas, whose semi-transverse axes will be respectively AC, OC, and whose semi-conjugate axes will be respectively OC, AC, and which therefore will be *conjugate* to each other.

CONIC SECTIONS. 121

70. As long as the cutting plane remains parallel to itself, it is evident that the ratio of AE to AX, and therefore of AS to AX will be altered. Hence the sections made by planes inclined at the same angle to the axis of the cone will have the same eccentricity.*

71. Through any point Q on the circle EQe let a plane be drawn parallel to ANP, intersecting the plane of the paper in the straight line WLN', the cone in the curve WQP', and the planes of the circular sections EQe, RPr in the ordinates QL, $P'N'$.

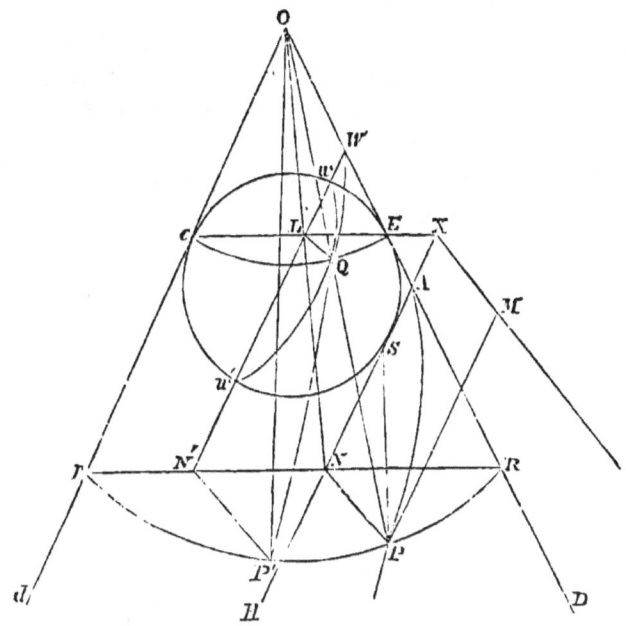

Then it is manifest that the curve WQP' will touch the circle wQw', formed by the intersection of the cutting plane with the sphere at the extremities of the ordinate QL produced.

Join OP' meeting EQe in Q'; then
$$P'Q' : N'L :: RE : NX,$$
$$:: SA : AX.$$

* The ratio of $SA : AX$, or of $CS : CA$, is called the eccentricity.

But $P'Q'$ is equal to the tangent drawn from P' to the circle wQw', and $N'L$ is equal to the perpendicular from P' on the common ordinate of the circle wQw' and the section WQP'.

Hence we have the following important property,—

If a circle touch a conic section in two points at the extremities of an ordinate, the ratio which the tangent drawn to the circle from any point on the curve bears to the perpendicular from the same point on the common chord is equal to the eccentricity of the conic section.

If the two points in which the circle touches the conic coincide, the circle becomes the circle of curvature at the vertex; and therefore the ratio which the tangent drawn from any point of a conic section to the circle of curvature at the vertex bears to the abscissa of the point, is constant and equal to the eccentricity of the curve.

PROBLEMS ON THE SECTIONS OF THE CONE.

1. The foci of all parabolic sections which can be cut from a given right cone lie upon the surface of another cone.

2. The foci of all elliptical sections of a given right cone, in which the ratio of CA to CS is the same, will lie on two other cones.

3. The extremities of the minor axes of the elliptical sections of a right cone made by parallel planes lie on two generating lines.

4. The latus rectum of a parabola cut from a given cone varies as the distance between the vertices of the cone and the parabola.

5. Under what conditions is it possible to cut an equilateral hyperbola from a given right cone?

6. Two cones whose vertical angles are supplementary are joined as in *Art.* 69, *Cor.* 2. Prove that the latera recta of the curves of section of the cones, whose axes are respectively OI and Oi, made by planes parallel or perpendicular to the plane of the axes, are in the duplicate ratio of Oi and OI.

ADDITIONAL PROBLEMS.

1. Show that the part of the directrix of a parabola, intercepted between the perpendiculars on it from the extremities of any focal chord, subtends a right angle at the focus.

2. The locus of the foci of all parabolas touching the three sides of a triangle is a circle. Prove this, and give a geometrical construction for finding the centre.

3. A system of parabolas which always touch two given straight lines have their axes parallel; show that the locus of the foci is a straight line.

4. Prove that the locus at the foot of the perpendicular from the focus of a parabola on the normal is a parabola.

5. If S be the focus of a parabola, which touches the sides AB, AC of the triangle ABC at the points B, C, and O the centre of the circle described about the triangle; prove that the angle OSA is a right angle.

6. From the focus of a parabola a straight line is drawn parallel to the tangent at any point P, meeting the diameter through P in V'; show that the tangent drawn from P to any circle passing through V' is equal to one-half of the ordinate QV, V being the second point in which the circle cuts the diameter through P.

7. PSp is a focal chord, and upon PS and pS as diameters, circles are described; prove that the length of either of their common tangents is a mean proportional between AS and Pp.

8. If AQ be a chord of a parabola through the vertex A, and QR be drawn perpendicular to AQ to meet the axis in R; prove that AR will be equal to the chord through the focus parallel to AQ.

9. The locus of the vertices of all parabolas, which have a common focus and a common tangent, is a circle.

10. Two parabolas have a common axis and vertex, and their concavities turned in opposite directions; the latus rectum of one is eight times that of the other; prove that the portion of a tangent to the former, intercepted between the common tangent and axis, is bisected by the latter.

11. B is a point on a radius OA of a circle, whose centre is O. On OA produced a point C is taken, such that $OB \cdot OC = OA^2$. If P be any point on the circumference of this circle, R the middle point of BP, and Q the point of intersection of AR, CP; prove that the locus of Q is a circle.

12. If from the middle point of a focal chord of a parabola two straight lines be drawn, one perpendicular to the chord meeting the axis in G, and the other perpendicular to the axis meeting it in N; show that NG is constant.

13. A circle is drawn touching the axis of a parabola, the focal distance of a point P, and the diameter through P. Show that the locus of its centre is a parabola with vertex S, and latus rectum equal to AS.

14. If from the point of intersection of the directrix and axis of a parabola a chord XPQ be drawn, cutting the parabola in P, Q; show that the rectangle contained by the ordinates of PQ is equal to the square of one-half the latus rectum.

15. Find the locus of the centre of a circle which touches a given circle and a given straight line.

16. Given one point of contact of a parabola with three

tangents given in position, find the two other points of contact.

17. The triangle ABC circumscribes a parabola whose focus is S. Through A, B, C, lines are drawn perpendicular respectively to SA, SB, SC. Show that these lines pass through one point.

18. From the focus a line is drawn parallel to the tangent at P, meeting the parabola at Q. QN is an ordinate, and the tangents at P and Q meet the axis in T and T'. Prove that $SN^2 = 4AT \cdot AT''$, and that if the diameter at P meet SQ in E, the locus of E is a parabola, whose latus rectum is half that of the given parabola.

19. P is any point in a parabola; through S a line is drawn at right angles to the axis, meeting the chord AP or AP produced in R. Prove that $SK \cdot SR = 2AS \cdot SY$, where SY is the perpendicular on the tangent, and SK on the normal.

20. From the focus S of a parabola SK is drawn, making a given angle with the tangent at P. Show that the locus of K is that tangent to the parabola which makes with the axis an angle equal to the given angle.

21. PSQ is a focal chord of a parabola, AP' a parallel chord meeting the latus rectum in Q'; prove that $AP' \cdot AQ = SP \cdot SQ$.

22. The circle of curvature at any point of a parabola whose abscissa is AN cuts the axis in U and U'. Prove that $AU \cdot AU' = 3AN^2$.

23. AB is a diameter of a circle. From any point Q in the circumference a tangent QP is drawn, and from P a perpendicular PN is let fall upon AB. Show that if P be always taken so that QP is equal to AN, the locus of P will be a parabola.

24. If a tangent be drawn from any point of a parabola to the circle of curvature at the vertex, the length of the tangent

will be equal to the abscissa of the point measured along the axis.

25. To two parabolas which have a common focus and axis two tangents are drawn at right angles; the locus of their intersection is a straight line parallel to the directrices.

26. If any three tangents be drawn to a parabola, the circle described about the triangle so formed will pass through the focus, and the perpendiculars from the angles on the opposite sides intersect in the directrix.

27. A parabola touches one side of a triangle in its middle point, and the other two sides produced. Prove that the perpendiculars drawn from the angles of the triangle upon any tangent to the parabola are in harmonical progression.

28. Two equal parabolas have the same axis and vertex, but are turned in opposite directions; chords of one parabola are tangents to the other. Show that the locus of the middle point of the chords is a parabola whose latus rectum is one-third of that of the given parabola.

29. Two equal parabolas have the same focus, and their axes are at right angles to each other, and a normal to one of them is perpendicular to a normal of the other; prove that the locus of the intersection of such lines is a parabola.

30. Show that in every ellipse there are two equal conjugate diameters, coinciding in direction with the diagonals of the rectangle, which touches the ellipse at the extremities of the axes.

31. If a circle be described through the two foci of an ellipse, cutting the ellipse, show that the angle between the tangents to this circle, and to the ellipse at either point of intersection, is equal to the inclination of the normal to the ellipse to the axis minor.

32. The points in which the tangents at the extremities of the transverse axis of an ellipse are cut by the tangent at any

point of the curve, are joined one with each focus; prove that the point of intersection of the joining lines lies in the normal at the point.

33. The external angle between any two tangents to an ellipse is equal to the semi-sum of the angles which the chord joining the points of contact subtends at the foci.

34. The tangent to an ellipse at any point P is cut by any two conjugate diameters in T, t, and the points T, t, are joined with the foci S, S' respectively; prove that the triangles $SPT, S'Pt$ are similar to each other.

35. P is any point on a fixed circle, the centre of which is O; E is a fixed point without the circle; an ellipse is described with centre O and area constant so as always to touch EP at P. Find the locus of the extremities of the diameter conjugate to OP.

36. The normal at any point P of an ellipse cuts the axes in G, g; prove that if any circle be described passing through G, g, the tangent to it from P is equal to CD.

37. Given a focus, a tangent, and the eccentricity of a conic section; prove that the locus of the centre is a circle.

38. A straight line is drawn through a given point C within a circle to cut it in P, P'. If p is taken in it such that $Cp^2 = CP \cdot CP'$, find the locus of p.

39. In the ellipse $PY \cdot PY' : PN^2 :: CS^2 : BC^2$ and $SY \cdot CD = SP \cdot BC$.

40. Show that if the distance between the foci of the ellipse be greater than the length of its axis minor, there will be four positions of the tangent, for which the area of the triangle, included between it and the straight lines drawn from the centre of the curve to the feet of the perpendiculars from the foci on the tangent, will be the greatest possible.

41. Two conjugate diameters of an ellipse are cut by the tangent at any point P in M, N; prove that the area of the triangle CPM varies inversely as that of the triangle CPN.

42. Circles are described on SY, $S'Y'$ as diameters, cutting SP, $S'P$ respectively in Q, Q'. Prove that $SQ \cdot S'P = SP \cdot S'Q' = BC^2$.

43. PSP', pSp' are any two focal chords of a conic section, P and p being on the same side of the axis; prove that Pp, $P'p'$ meet on the directrix.

44. Prove that an ellipse can be inscribed in any parallelogram so as to touch the middle points of the four sides; and show that this ellipse is the greatest of all inscribed ellipses.

45. If from any point on the exterior of two concentric, similar, and similarly placed ellipses, two tangents be drawn to the interior ellipse which also intersect the exterior; show that the distance between the points of intersection will be double of that between the points of contact.

46. The tangent at any point P in an ellipse, of which S and H are the foci, meets the axis major in T, and TQR bisects HP in Q, and meets SP in R; prove that PR is one-fourth of the chord of curvature at P through S.

47. Prove that the distance between the two points on the circumference of an ellipse at which a given chord, not passing through the centre, subtends the greatest and least angles, is equal to the diameter which bisects that chord.

48. From any point on the auxiliary circle chords are drawn through the foci of an ellipse, and straight lines join the extremities of the chords with the extremity of the diameter passing through the point; prove that these lines will touch the ellipse.

49. A quadrilateral circumscribes an ellipse. Prove that either pair of opposite sides subtends supplementary angles at either focus.

50. Two tangents to an ellipse intersect at right angles; show that the straight line joining their point of intersection with the point of intersection of the normals at the points of contact passes through the centre.

51. P, Q are points in two confocal ellipses, at which the line joining the common foci subtends equal angles; prove that the tangents at P, Q are inclined to an angle which is equal to the angle subtended by PQ at either focus.

52. Tangents to an ellipse are drawn from any point in a circle through the foci; prove that the lines bisecting the angle between the tangents all pass through a fixed point.

53. If the ordinate at P meet the auxiliary circle in Q, and CQ meet the ellipse in R, then CR is equal to the perpendicular on the tangent at P from C.

54. If P be a point such that SP, $S'P$ are perpendicular; prove that $CD^2 = 2 \cdot BC^2$.

55. If circles be described to the triangle SPS' opposite to the angles S and S'; prove that the rectangle contained by their radii is equal to BC^2.

56. The circle of curvature at any point P of an ellipse meets the focal distances in R, R'; SU is a tangent to the circle.

Prove that $SU^2 : SP^2 :: 2 \cdot SP - 3 \cdot AC : AC$,

and if RR' passes through the centre of the circle of curvature, $CP = CS$. Determine the limits of possibility in both cases.

57. A straight line is drawn from the centre of an ellipse meeting the ellipse in P, the circle on the major axis in Q, and the tangent at the vertex in T. Prove that as CT approaches and ultimately coincides with the semi-major axis, TP and QT are ultimately in the duplicate ratio of the axes.

58. A straight line is drawn through the focus S of an ellipse meeting the two tangents at right angles to it in Y and Z, the diameter parallel to these tangents in L, and the directrix in M; prove that

$$SL : SY :: SZ : SM.$$

59. If any equilateral triangle PQR be described in the auxiliary circle of an ellipse, and the ordinates to P, Q, R meet the ellipse in P', Q', R'; the circles of curvature at P', Q', R', meet in one point lying on the ellipse.

60. From a point T two tangents TP, TQ are drawn to an ellipse. Show that a circle with T as centre can be described so as to touch $SP, S'P, SQ, S'Q$.

61. If the normal at P meet the axis minor in g, and if the tangent at P meet the tangent at the vertex A in V; show that

$$Sg : SC :: PV : VA.$$

62. If a circle passing through Y and Y' touch the major axis in Q, and that diameter of the circle which passes through Q meet the tangent in P; show that $PR = BC$. (See fig. Prop. XV.)

63. If PG the normal at P cut the major axis in G, and if DR, PN be the ordinates of D and P, prove that the triangles PGN, DRC are similar; and thence deduce that PG bears a constant ratio to CD.

64. The tangent at a point P of an ellipse meets the tangents at the vertices in V, V'. On VV' as diameter, a circle is described which intersects the ellipse in Q, Q'; show that the ordinate of Q is to the ordinate of P as BC to $BC + CD$ where CD is conjugate to CP.

65. PCP' is any diameter of an ellipse; the tangents at any two points E and E' intersect in F; $PE', P'E$ intersect in G. Show that FG is parallel to the diameter conjugate to PCP'.

66. If P be any point on an ellipse, and with P as centre and the semi-axis minor as radius a circle be described; prove that if PG be the normal, a circle described on CG as diameter will cut the first circle at right angles.

67. ABC is an isosceles triangle having $AB = AC$. BD, BE drawn on opposite sides of BC, and equally inclined to it, meet AC in D and E.

If an ellipse be described about BDE having its minor axis parallel to BC; then AB will be a tangent to the ellipse.

68. If AQ be drawn from one of the vertices of an ellipse perpendicular to the tangent at any point P; prove that the locus of the point of intersection of PS and QA produced will be a circle.

69. If Y, Y' be the feet of the perpendiculars from the foci of an ellipse on the tangent at P; prove that the circle circumscribed about the triangle YNY' will pass through C.

70. Prove that the angle between the tangents to the auxiliary circle at Y, Y' is the supplement of the angle SPS'.

71. P is any point on an ellipse; PM, PN perpendicular to the axes meet respectively, when produced, the circles described on the axes as diameter in the points Q, Q'. Show that QQ' passes through the centre.

72. Assuming that the greatest triangle which can be inscribed in a circle is equilateral, prove by the method of projection, that the greatest triangle which can be inscribed in an ellipse has one of its sides bisected by a diameter of the ellipse, and the others cut in points of bisection by the conjugate diameter.

73. PQ is a chord of an ellipse, normal at P, LCL' the diameter bisecting it. Show that PQ bisects the angle LPL' and that $LP + L'P$ is constant.

74. A tangent to an ellipse at a point P intersects a fixed tangent in T; if through the focus a line be drawn perpendicular to ST meeting the tangent to P in Q; show that the locus of Q is a straight line touching the ellipse.

75. In an ellipse if a line be drawn through the focus making a constant angle with the tangent; prove that the locus of the point of intersection with the tangent is a circle.

76. Any chord PP' of an ellipse is produced to a point Q, such that $P'Q$ is equal to half the diameter parallel to PP', and $QR'R$ is drawn through the centre to meet the ellipse in R, R'; show that the area PCR is three times the area $P'CR'$.

77. In an ellipse, L is the extremity of the latus rectum, and CD conjugate to CL. If a circle be described with centre C and passing through D, and a line be drawn through D parallel to the major axis, the portion of this line which lies within the circle will be equal to the latus rectum.

78. If P be any point in an ellipse, and K the point in which a normal at P intersects a line at right angles to it through S', E the point of intersection of SP, and the diameter conjugate to CP, and if EK and CK be joined, each of the figures $SCKE$, $S'CEK$ will be a parallelogram.

79. If T be a point on the axis AA' produced, and PN the ordinate of the point where the tangent from T touches the ellipse; prove that

$$AN \cdot A'N : AT \cdot A'T :: CN : CT.$$

80. Given in an ellipse a focus and two tangents; prove that the locus of the other focus is a straight line.

81. A focus, a tangent, and the axis major being given, prove that the locus of the other focus is a circle.

82. A focus, a tangent, and the axis minor being given, prove that the locus of the other focus is a straight line.

83. An ellipse touches a fixed ellipse and has a common focus with it; if the major axis be fixed, the locus of the other focus is a circle; if the minor axis be fixed, the locus is an ellipse.

84. An ellipse and a parabola have a common focus. Prove that the ellipse either intersects the parabola in two points, and has two common tangents with it, or else does not cut it.

85. If in the ellipse a focus, a point, and the axis minor be given, the locus of the other focus is a parabola.

86. If at the extremities P, Q of any two diameters CP, CQ of an ellipse, two tangents Pp, Qp be drawn cutting each other in T, and the diameter produced in p and q, then the area of the triangles TQp, TPq are equal.

87. If a straight line CN be drawn from the centre to bisect that chord of the circle of curvature at any point P of an ellipse, which is common to the ellipse and circle, and if it be produced to cut the ellipse in Q, and the tangent in T; prove that $CP = CQ$, and that each is a mean proportional between CN and CT.

88. An ellipse is described so as to touch the three sides of a triangle; prove that if one of its foci move along the circumference of a circle passing through two of the angular points of the triangle, the other will move along the circumference of another circle, passing through the same two angular points. Prove also that if one of these circles pass through the centre of the circle inscribed in the triangle, the two circles will coincide.

89. A triangle is described about an ellipse, so that the extremities of one of its sides lie in an ellipse, confocal with the given one; prove that the line bisecting the opposite angle passes through the pole of that side with respect to the outer ellipse.

90. Prove the following construction for a pair of tangents from any external point T to an ellipse of which the centre is C. Join CT; let $TPCP''T$, a similar and similarly situated ellipse, be drawn, of which CT is a diameter, and P, P'' its points of intersection, with the given ellipse; TP, TP'' will be tangents to the given ellipse.

91. The locus of the foci of all ellipses inscribed in the same parallelogram is a rectangular hyperbola. Prove this, and give a geometrical construction for finding the asymptotes.

92. AC is a fixed diameter of a circle, $ABCD$ a quadrilateral figure inscribed in the circle; prove that if the angles BAC, DAC be complementary, the locus of the intersection of BA, CD will be an hyperbola.

93. Prove that a circle can be described so as to touch the four straight lines drawn from the foci of an hyperbola to any two points on the same branch of the curve.

94. Any three diameters of an ellipse, LL', MM', NN', being taken, a circumscribing parallelogram $RTUV$ touches the ellipse at L, L', M, M'. Show that a conic section can be described through the points R, T, U, V, N, N', which will be an hyperbola whose asymptotes are the lines forming in the ellipse the diameters conjugate to NN' and to the other common chord of the ellipse and hyperbola.

95. On opposite angles of any chord of a rectangular hyperbola are described equal segments of circles; show that the four points in which the circles to which these segments belong again meet the hyperbola, are the angular points of a parallelogram.

96. A triangle is inscribed in a rectangular hyperbola: prove that the circle described through the middle points of the sides of the triangle passes through the centre of the hyperbola.

97. ACB is an isosceles triangle; AB the base, and D any point in CB or CB produced: if BZ be drawn parallel to AD, meeting CA or CA produced in Z, prove that the middle point of DZ will be in an hyperbola whose asymptotes are CA, CB.

98. An ellipse and hyperbola are described so that the foci of each are at the extremities of the transverse axis of the other; prove that the tangents at their points of intersection meet the conjugate axis in points equidistant from the centre.

99. In a rectangular hyperbola, PK, PL are drawn at right angles to $A'P$, AP respectively to meet the transverse axis in K and L; prove that PK is equal to AP and KL to AA', and the normal at P bisects KL.

100. In a rectangular hyperbola PC is a fixed diameter, Q any point on the curve; show that the angles QPC, QCP differ by a constant angle.

101. If the tangent at any point P of an hyperbola cut an asymptote in T, and if SP cut the same asymptote in Q, then $SQ = QT$.

102. If a given point be the focus of any hyperbola, passing through a given point and touching a given straight line, prove that the locus of the other focus is an arc of a fixed hyperbola.

103. At any P of an hyperbola a tangent is drawn, and PQ is taken on it in a constant ratio to CD; prove that the locus of Q is an hyperbola.

104. In an hyperbola, supposing the two asymptotes and one point of the curve be given in position, show how to construct the curve; and find the position of the foci.

105. If A, D be two fixed points, and the angle PAD always exceed PDA by a given angle; find the locus of P, and the position of the transverse axis and asymptote.

106. From the middle point D of the base AB of the triangle ABC a straight line EDE' is drawn, making a given angle with AB, and the points E, E' are taken so that $ED = E'D = \frac{1}{2} AB$. If CA, CB take all possible positions consistent with the condition that the difference of the angles CAB, CBA is equal to EDA; prove that the point C will trace out a rectangular hyperbola of which AB, $E'E$ are conjugate diameters.

107. In the rectangular hyperbola, prove that the triangle, formed by the tangent at any point and its intercepts on the axes, is similar to the triangle formed by the straight line joining that point with the centre, and the abscissa and semi-ordinate of the point.

108. Tangents are drawn to an hyperbola, and the portion of each tangent intercepted by the asymptotes is divided in a constant ratio; prove that the locus of the point of section is an hyperbola.

109. Show that the point of trisection of a series of conterminous circular arcs lie on branches of two hyperbolas, and determine the distance between their centres.

110. From a point R on one asymptote RE is drawn touching the hyperbola in E, and ET, EV are drawn through E, parallel to the asymptotes, cutting a diameter in T and V; RV is joined, cutting the hyperbola in P, p: show that TP, Tp touch the hyperbola.

111. Given in the ellipse a focus and two points, the locus of the other focus is an hyperbola.

112. If a rectangular hyperbola passes through three given points, the locus of its centre is a circle, which passes through the middle points of the lines joining the three given points.

113. If the tangent at P meet one asymptote in T, and a line TQ be drawn parallel to the other asymptote to meet the curve in Q; prove that if PQ be joined and produced both ways to meet the asymptotes in R and R', RR' will be trisected at the points P and Q.

114. If two concentric rectangular hyperbolas have a common tangent, the lines joining their points of intersection to their respective points of contact with the common tangent will subtend equal angles at their common centres.

115. If TP, TQ be two tangents drawn from any point T to touch a conic in P and Q, and if S and S' be the foci, then
$$ST^2 : S'T^2 :: SP \cdot SQ : S'P \cdot S'Q.$$

116. The circle of curvature at the vertex of a conic meets the axis again in D, and a tangent is drawn to the circle at D; if two tangents be drawn to the circle from any point in the conic they will intercept between them a constant length of the former tangent.

117. If the lines which bisect the angles between pairs of tangents to an ellipse be parallel to a fixed straight line, prove that the locus of the points of intersection of the tangents will be a rectangular hyperbola.

118. An hyperbola, of given eccentricity, always passes through two given points; if one of its asymptotes always pass through a third given point in the same straight line with these, prove that the locus of the centre of the hyperbola will be a circle.

119. A, P and B, Q are points taken respectively in two parallel straight lines, A and B being fixed, and P, Q variable. Prove that if the rectangle $APBQ$ be constant, the line PQ will always touch a fixed ellipse or a fixed hyperbola, according as P and Q are on the same or opposite sides of AB.

120. If two plane sections of a right cone be taken, having the same directrix, the foci corresponding to that directrix lie on a straight line which passes through the vertex.

121. Give a geometrical construction by which a cone may be cut so that the section may be an ellipse of given eccentricity.

122. Given a right cone and a point within it, there are but two sections which have this point for focus; and the planes of these sections make equal angles with the straight line joining the given point and the vertex of the cone.

123. If the curve formed by the intersection of any plane with a cone be projected upon a plane perpendicular to the axis, prove that the curve of projection will be a conic section having its focus at the point in which the axis meets the plane of projection.

124. If F be the point where the major axis of an elliptic section meets the axis of the cone, and C be the centre of the section; prove that
$$CF : CS :: AA' : AO + A'O,$$
O being the vertex of the cone.

SECOND SERIES.

1. If OQ, OQ' be tangents to a parabola, and OV drawn parallel to the axis meet the directrix in K and QQ' in V, and QQ' meet the axis in N, $OKNS$ shall be a parallelogram.

2. G is the foot of a normal at a point P of a parabola, Q is the middle point of SG, and X is the foot of the directrix, prove that
$$QX^2 - QP^2 = 4AS^2.$$

3. Through any point O tangents are drawn to a parabola; and through O straight lines are drawn parallel to the normals at the points of contact, prove that one diagonal of the parallelogram so formed passes through the focus.

4. Two parabolas have their foci coincident, prove that the common chord passes through the intersection of the directrices, and that they cut one another at angles which are half the angles between the axes.

5. A circle is drawn through the point of intersection of two given straight lines and through another given point, prove that the straight line joining the points, where the circle again meets the two given straight lines, touches a fixed parabola.

6. Common tangents are drawn to two parabolas which have a common directrix and intersect in P, Q; prove that the chords joining the points of contact in each parabola are parallel to PQ; and the part of each tangent between its points of contact with the two curves is bisected by PQ produced.

7. The tangents at two points Q, Q' in the parabola meet the tangent at P in R, R' respectively, and the diameter through their point of intersection T meets it in K; prove that $PR = KR'$ and that if QM, $Q'M$,$'$ TN be the ordinates of Q, Q,$'$ T' respectively to the diameter through P, PN is a mean proportional between PM and PM'.

8. P and p are points on a parabola on the same side of the axis and PN, pn the ordinates; the normals at P and p intersect in Q; prove that the distance of Q from the axis

$$= \frac{2PN \cdot pn(PN + pn)}{(\text{latus rectum})^2}.$$

Deduce the value of the radius of curvature.

9. Two equal parabolas have a common focus, and from any point on the common tangent another tangent is drawn to each; prove that these tangents are equidistant from the common focus.

10. From an external point O two tangents are drawn to a parabola, and from the points where they meet the directrix two other tangents are drawn meeting the tangents from O at A and B.

Prove that AB passes through the focus S, and that OS is at right angles to AB.

11. Given two tangent lines to a parabola and the focus, show how to determine the curve.

12. Inscribe in a given parabola a triangle having its sides parallel to those of a given triangle.

13. Normals at P, p the extremities of a focal chord of a parabola meet the axis in G and g. The tangents meet in T. Prove that the circles about the triangles SPG, Spg intersect on TS produced at a point R such that S bisects RT.

14. A circle and parabola touch one another at both ends of a double ordinate to the parabola, prove that the latus rectum is a third proportional to the parts into which the abscissa of the points of contact is divided by the circle either internally or externally.

15. PQ is a chord of a parabola normal at P; QR is drawn parallel to the axis to meet the double ordinate PP' produced in R; then the rectangle contained by PP' and $P'R$ is constant.

16. OP, OQ are two tangents to a parabola, and OM is drawn perpendicular to the axis; if R be the point where PQ cuts the axis, MR is bisected by the vertex.

17. A parabola, whose focus is S, touches the three sides of a triangle ABC, bisecting the base BC in D; prove that AS is a fourth proportional to AD, AB, and AC.

18. If the chord of contact be normal at one end, the tangent at the other is bisected by the perpendicular through the focus to the line joining the focus to the external point.

19. Two parabolas have a common focus, and from any point on their common tangent are drawn two other tangents to the parabolas; prove that the angle between them is equal to the angle between the axes of the parabolas.

20. If tangents be let fall on any tangent to a parabola from two given points on the axis equidistant from the focus, the difference of their squares is constant.

21. If a quadrilateral be inscribed in a circle, one of the three diagonals of the quadrilateral passes through the focus of the parabola which touches its four sides.

22. A chord of a parabola is drawn parallel to a given straight line, and on this chord as diameter a circle is described; prove that the distance between the middle point of this chord, and of the chord joining the other two points of intersection of the circle and parabola will be of constant length.

23. Through any point P of an ellipse a line is drawn perpendicular to the radius vector CP meeting the auxiliary circle in R, R': prove that RR' is equal to the difference of the focal distances of the extremity of the diameter conjugate to CP.

24. An ellipse and parabola whose axes are parallel, have the same curvature at a point P and cut one another in Q; if the tangent at P meet the axis of the parabola in T prove that PQ is four times PT.

25. A triangle is inscribed in an ellipse so that each side is parallel to the tangent at the opposite angle: prove that the sum of the squares on the sides is to the sum of the squares on the axes as nine to eight.

26. In an ellipse the perimeter of the quadrilateral formed by the tangent, the perpendiculars from the foci and the transverse axis, will be the greatest possible, when the focal distances of the point of contact are at right angles to each other.

27. Two given ellipses in the same plane have a common focus, and one revolves about the common focus, while the other remains fixed; prove that the locus of the point of intersection of their common tangents is a circle.

28. Perpendiculars $SY, S'Y'$ are drawn from the foci upon a pair of tangents TY, TY'; prove that the angles STY, $S'TY'$ are equal or supplementary to the angles at the base of the triangle formed by joining Y, Y' to the centre of the ellipse.

29. With the centre of perpendiculars of a triangle as centre, are described two ellipses, one about the triangle, and the other touching its sides: prove that these ellipses are similar and their homologous axes at right angles.

30. Through a point P of an ellipse a line PDE is drawn cutting the axes so that the segments PD and PE are equal to the two semi-axes respectively; perpendiculars to the axes through D and E intersect in O; prove that PO is a normal.

31. If the focal distance SP of an ellipse meet the conjugate diameter in E; then the difference of the squares on CP and SE will be constant.

32. The chords of curvature through any two points of an ellipse in the direction of the line joining them are in the same ratio as the squares of the diameters parallel to the tangents at the points.

33. From the extremities of the diameter of an ellipse, perpendicular to one of the equi-conjugate diameters, chords are drawn parallel to the other. Prove that these chords are normal to the ellipse.

34. An ellipse of given semi-axes touches three sides of a given rectangle; find its centre and foci.

35. If P, Q be any two points on a fixed ellipse, whose foci are S, H and if SP, HQ intersect within the ellipse at R, prove that two ellipses can be drawn touching each other at R, the one having S for focus and touching the given ellipse at Q, the other having H for focus and touching the given ellipse at P.

If the major axis of one of the variable ellipses be given find the loci of their other foci, and of their point of contact.

36. Find the positions of the foci and directrices of an ellipse, which touches at two given points P, Q, two given straight lines PO, QO, and has one focus on the line PQ, the angle POQ being less than a right angle.

37. If Pp be drawn parallel to the transverse axis of a given ellipse, meeting the ellipse in P, p, and the circle whose diameter is the conjugate axis in R, r, then shall

$$Pp : Rr :: QQ' : PP'.$$

38. Through any point P of an ellipse are drawn straight lines APQ; $A'PR$ meeting the auxiliary circle in Q, R, and ordinates Qq, Rr are drawn to the transverse axis; prove that, L being an extremity of the latus rectum

$$Aq \cdot A'r : Ar \cdot A'q :: AC^2 : SL^2.$$

39. Two ellipses whose axes are equal, each to each, are placed in the same plane with their centres coincident, and axes inclined to each other. Draw their common tangents.

40. If S, S' be the two foci of an ellipse; and SY the perpendicular from S upon any tangent, prove that $S'Y$ will bisect the corresponding normal.

41. OR is a diagonal of the parallelogram of which OQ, OQ', tangents to an ellipse, are adjacent sides: prove that if R be in the ellipse, O will lie on a similar and similarly situated concentric ellipse.

42. A circle passes through a focus of an ellipse, has its centre on the major axis of the ellipse, and touches the ellipse: shew that the straight line from the focus to the point of contact is equal to the latus rectum.

43. If the focus of an ellipse be joined with the point where the tangent at the nearer vertex intersects any other tangent, and perpendiculars be drawn from the other focus on the joining line, and the last mentioned tangent, prove that the distance between the feet of these perpendiculars is equal to the distance from either focus to the remoter vertex.

44. A parallelogram is described about an ellipse; if two of its angular points lie on the directrices, the other two will lie on the auxiliary circle.

45. PS is the focal chord of a point on an ellipse; CR is a radius of the auxiliary circle parallel to PS, and drawn in the direction from P to S; SQ is a perpendicular on CR; shew that the rectangle contained by SP and QR is equal to the square on half the minor axis.

46. From the extremity P of the diameter PQ of an ellipse the tangent TPT' is drawn meeting two conjugate diameters in T, T'. From PQ the lines PR, QR are drawn parallel to the same conjugate diameters. Prove that the rectangle under the semi-axes of the ellipse is a mean proportional between the triangles PQR and CTT'.

47. If CP and CD be equal conjugate diameters of an ellipse, and the tangent and normal at P meet the major axis in T and G respectively, prove that $TC \cdot TG = 2CP^2$.

48. An ellipse is inscribed in a triangle having its centre at the centre of the circumscribed circle of the triangle; prove that the perpendiculars from the corners of the triangle on the opposite sides will be normals to the ellipse.

49. Tangents are drawn from any point on the auxiliary circle to the ellipse; prove that the line joining one of the points of contact with one of the foci is parallel to the line joining the other point of contact with the other focus.

50. From a point P without an ellipse, PQ is drawn parallel to the major axis, Q being either of the points in which it meets the curve; then the straight line bisecting PQ at right angles, the tangent at Q, and the line which joins the middle points of PK, PL (the tangents drawn from P) meet in a point.

51. CP and CD are conjugate semi-diameters of an ellipse; PQ is a chord parallel to one of the axes; shew that DQ is parallel to one of the straight lines which join the ends of the axes.

52. A series of ellipses pass through the same point and have a common focus and their major axes of the same length; prove that the locus of their centres is a circle.

What are the limits of the eccentricities of the ellipses?

53. A parallelogram is inscribed in an ellipse, and from any point on the ellipse two straight lines are drawn parallel to the sides of the parallelogram; prove that the rectangles under the segments of these straight lines, made by the sides of the parallelogram, will be to one another in a constant ratio.

54. If the tangent and normal in an ellipse meet the axis in T and G respectively, and Q be the corresponding point of the auxiliary circle, then

$$TQ : TP :: BC : PG.$$

55. PCP' is a diameter of an ellipse, CD conjugate to CP; prove that PD, DP' are inversely proportional to the diameters which bisect them.

56. If EF be one side of a parallelogram described about an ellipse, having its sides parallel to conjugate diameters, and the lines joining E, F to the foci intersect in O, O', shew that O, O' and the foci will lie on a circle.

57. If the normal at any point of a hyperbola meet the conjugate axis in g, and S be the focus, Pg will be to Sg in a constant ratio.

58. In the rectangular hyperbola, if circles be described passing through the centre of the hyperbola and the points of intersection of the tangent with the pairs of conjugate diameters, their centres will lie on a fixed straight line.

59. A chord of a rectangular hyperbola is drawn perpendicular to a fixed diameter and the extremities of the chord and diameter are joined by four straight lines; then the line joining the two fresh points of intersection of these lines will be parallel to a fixed line.

60. A series of confocal ellipses is cut by a confocal hyperbola; prove that either focal distance of any point of intersection is cut by its conjugate diameter with respect to the particular ellipse in a point which lies on a circle.

61. If from any point on the hyperbola a tangent be drawn to the circle described on the transverse axis as diameter, its length is equal to the semi-minor axis of the confocal ellipse through the point.

62. With one focus of a given hyperbola as focus, and any tangent to the hyperbola as directrix, is described another hyperbola touching the conjugate axis of the former; prove that the two will be similar.

63. Two tangents are drawn to the same branch of a rectangular hyperbola from an external point; prove that the angles which these tangents subtend at the centre are respectively equal to the angles which they make with the chord of contact.

64. A circle is described through P, P', the extremities of any diameter of a rectangular hyperbola, and cutting the tangent at P and T; prove that $P'T$ and the tangent to the circle at P meet on the hyperbola.

65. A fixed hyperbola is touched by a concentric ellipse. If the curvatures at the point of contact are equal, the area of the ellipse will be constant.

66. Two equal circles touch a rectangular hyperbola at a point O, and intersect it again in P, Q; P', Q' respectively; prove that the points may be so taken that PP', QQ' each subtend a right angle at O, and that the straight lines joining P', Q' to the centre of the circle OPQ will trisect OP, OQ respectively.

67. In a central conic given a focus, the length of the transverse axis, and that the second focus lies on a fixed straight line; prove that the conic will touch two fixed parabolas having the given focus for focus.

68. In a hyperbola a circle is described about a focus as centre, with radius one-fourth of the latus rectum; prove that the focal distances of the points of intersection with the hyperbola are parallel to the asymptotes.

69. Prove that the angles subtended at the vertices of a rectangular hyperbola by any chord parallel to the conjugate axis are supplementary.

70. A line moves in such a manner that the sum of the spaces on its distances from two given points is constant; prove that it always touches an ellipse or hyperbola, the square on whose transverse axis is equal to twice the sum of the squares on the distances of the moving line from the given points.

71. The eccentricity of a hyperbola is 2. A point D is taken on the axis, so that its distance from the focus S is equal to the distance of S from the further vertex A', P being any point on the curve, $A'P$ meets the latus rectum in K. Prove that DK and SP intersect on a certain fixed circle.

72. In a hyperbola if PH, PK be drawn parallel to the asymptotes, and a line through the centre meet PH, PK in R, T, and the parallelogram $PRQT$ be completed, Q is a point on the hyperbola.

73. If an ellipse and hyperbola have their axes coincident and proportional, points on them equidistant from one axis have the sum of the squares on their distances from the other axis constant.

74. In the hyperbola prove that (*see fig. Prop. XXIII.*)
$$MD : PN :: AC : BC :: CN : CM.$$

75. A parabola and hyperbola have the same focus and directrix, and SPQ is a line through the focus S to meet the parabola in P, and the nearer branch of the hyperbola in Q; prove that PQ varies as the rectangle contained by SP and SQ.

76. If two hyperbolas have their transverse axes parallel, and their eccentricities equal, they will have parallel asymptotes. Does the converse hold?

77. PP' is any diameter of a rectangular hyperbola, Q any point on the curve, PR, $P'R'$ are drawn at right angles to PQ, $P'Q$ respectively, intersecting the normal at Q in R, R': prove that QR and QR' are equal.

78. In a hyperbola a line parallel to BC, through the intersection of the tangent at P with the asymptote, meets DP, CA produced in the same point, CD and CP being conjugate.

79. From a given point in a hyperbola draw a straight line such that the segment intercepted between the other intersection with the hyperbola and a given asymptote shall be equal to a given line. When does the problem become impossible?

80. A tangent is drawn from any point on the transverse axis of a hyperbola to the auxiliary circle; if the angle between this tangent and the ordinate be called the eccentric angle, shew that the eccentric angles at the points of intersection of an ellipse with two hyperbolas, confocal with itself, are equal.

81. Two unequal parabolas have a common focus and axes opposite; a rectangular hyperbola is described touching both parabolas and having its centre at the common focus; prove that the angle between the lines joining the points of contact to the common focus is $60°$.

82. The tangent and normal at any point of a hyperbola intersect the asymptotes and axes respectively in four points which lie on a circle passing through the centre of the hyperbola; and the radius of this circle varies inversely as the perpendicular from the centre upon the tangent.

83. Prove that the rectangular hyperbola which has for its foci the foci of an ellipse, will cut the ellipse at the extremities of the equal conjugate diameters.

84. ABC is a given triangle. CA, CB produced are the asymptotes of a hyperbola cutting AB in P. Find the position of P when the sum of the squares on its axes is the greatest possible.

85. From the point of intersection of an asymptote and a directrix of a hyperbola a tangent is drawn to the curve; prove that the line joining the point of contact with the focus is parallel to the asymptote.

86. Find the position and magnitude of the axes of a hyperbola which has a given line for asymptote, passes through a given point, and touches a given straight line in a given point.

87. Prove that a hyperbola can be described passing through the extremities of any two diameters of a given ellipse, having diameters conjugate to these for its asymptotes.

88. PQ is a normal at a point P of a rectangular hyperbola meeting the curve again in Q; prove that PQ is equal to the diameter of curvature at P.

89. P is a point on a hyperbola whose foci are S and H; another hyperbola is described whose foci are S and P, and whose transverse axis is equal to $SP - 2PH$; shew that the hyperbolas will meet only at one point, and that they will have the same tangent at that point.

90. The tangent at any point on a hyperbola is produced to meet the asymptotes, thus forming a triangle; determine the locus of the point of intersection of the straight lines drawn from the angles of this triangle to bisect the opposite sides.

91. The extremities of the latera recta of all conics which have a common major axis lie on two parabolas.

92. Having given three points, prove that there are four straight lines such, that with any one of them as directrix, and any one of the given points as focus, a conic section may be described passing through the other two given points.

93. In any conic section if PG, pg, the normals at the ends of a focal chord intersect in O, the straight line through O parallel to Pp bisects Gg.

94. The normal at P to a central conic meets the axes in G and g; GK and gk are perpendicular to the focal distance SP; then PK and Pk are constant.

If kl parallel to the transverse axes meet the normal at P in C, then kl will be constant.

95. A focal chord PSQ of a conic is produced to meet the directrix in K, and KM, KN are drawn through the feet of the ordinates PM, QN of P and Q.

If KN produced, meet PM produced, in R, prove that PR is equal to PM.

96. The normals at the extremities of a focal chord PSQ of a conic intersect in K, and KL is drawn perpendicular to PQ; KF is a diagonal of the parallelogram of which SK, SL are adjacent sides; prove that KF is parallel to the transverse axis of the conic.

97. With any point on a given circle as focus and a given diameter as directrix, is described a conic similar to a given conic; prove that all such conics will touch the two similar conics to which the given diameter is a latus rectum.

98. Every conic section passing through the centres of the four circles which touch the sides of a triangle, is a rectangular hyperbola ; and the locus of the centre of this system of rectangular hyperbolas is the circle circumscribing the triangle.

99. Find the locus of the centres of plane sections of a right cone drawn through a fixed point on the axis of the cone.

100. Shew how to cut a given cone, so that the section may be a parabola of given latus rectum.

101. If sections of a right curve be made, perpendicular to a given plane containing the axis, so that the distance between a focus of a section and that vertex which lies on the same generating line in the given plane be constant, prove that the transverse axes, produced if necessary, of all the sections will touch one of two fixed circles.

102. Find the position of the vertex and axis of a cone of given vertical angle, in order that a given parabola may be a section of the cone.

103. Two right cones have a common vertex; shew how to cut them by a plane so that the sections may be similar and similarly situated curves (1) when they intersect (2) when they do not intersect.

104. The vertex of a cone and the centre of a sphere inscribed within it, are given in position: a plane section of the cone at right angles to any generating line of the cone, touches the sphere; prove that the locus of the point of contact is a surface generated by the revolution of a circle, which touches the axis of the cone at the centre of the sphere.

105. What difference is there in the problems of cutting a given ellipse and a given hyperbola from a right cone?

An ellipse and hyperbola are so situated that the vertices of each curve are the foci of the other, and the curves are in planes at right angles to each other. If P be a point on the ellipse and Q a point on the hyperbola, S the focus, and A the interior focus of that branch of the hyperbola, then

$$PQ + AS = PS + AQ.$$

106. Two cones, whose vertical angles are supplementary, are placed with their vertices coincident and their axes at right angles, and are cut by a plane perpendicular to the common generating line; prove that the directrices of the section of one cone pass through the foci of the section of the other.

APPENDIX.

HARMONIC RATIO.

POLES AND POLARS.

RECIPROCATION.

72. DEF. If a straight line AB be divided internally in a point O, and externally in a point O' so that

$$AO : OB :: AO' : O'B$$

```
A ———— O ———— B ———————— O'
```

then AB is said to be *harmonically* divided in O and O'.

The point O' will evidently be on AB produced through the point B or A, according as AO is greater or less than OB.

If AO' is equal to OB, the point O' moves off to an infinite distance.

It is easy to see that the definition of Harmonic ratio given above coincides with the Algebraical Definition.

For if AO', OO', BO' be three quantities in Harmonic Progression, then by the Algebraical Definition

$$AO' : O'B :: AO' - OO' : OO' - BO'$$
$$:: AO : BO$$

N.B.—In Algebraical treatises 3 quantities are said to be in Harmonic Progression when the 1st is to the 3rd as the difference between the 1st and 2nd is to the difference between the 2nd and 3rd, from which definition it at once follows that the reciprocals of the quantities are in Arithmetical Progression.

Prop. I.

Having given the point O, to determine the point O',

Upon AB describe a segment of the circle containing *any* angle.

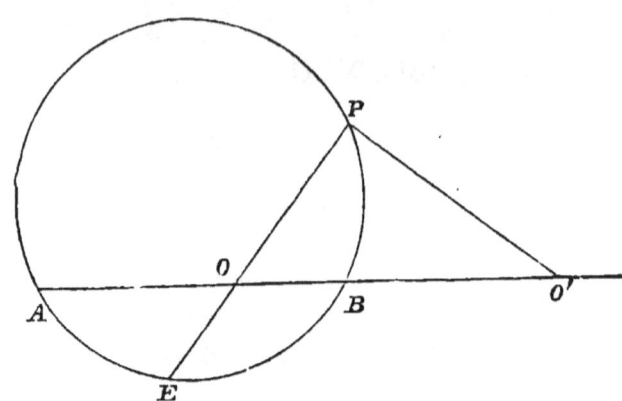

Bisect either portion of the circumference AB in the point E; join EO, and produce it to meet the other portion of the circumference in the point P.

Draw PO' at right angles to OP, meeting AB produced in O'; then

O' is the point required.

Since E is the middle point of the arc AB,

∴ the angle APB is bisected by PO,

∴ PO' bisects the *supplementary* angle formed by producing AP;

∴ $AO' : O'B :: AO : OB$. (*Euclid*, VI. A.)

Cor. If O' is given, then to determine the point O we have simply to divide AB in the ratio of the two given lines AO', $O'B$.

Prop. II.

If AB is *harmonically* divided in O and O'; then OO' is *harmonically* divided in B and A.

Since $AO : OB :: AO' : O'B$,

$\therefore AO : AO' :: OB : O'B$,

or $OB : BO' :: OA : AO'$,

$\therefore OO'$ is harmonically divided in B and A.

Prop. III.

If AB is harmonically divided in O and O', and AB be bisected in C; then

$$CO \cdot CO' = CB^2$$

$\overline{\ \ \ A\ \ \ \ \ \ \ \ \ \ \ C\ \ \ \ \ \ O\ \ \ \ \ B\ \ \ \ \ \ \ \ \ \ \ \ O'\ \ \ }$

Since $AO : OB :: AO' : O'B$,

$\therefore AO + OB : AO - OB :: AO' + O'B : AO' - O'B$;

or $2CB : 2CO :: 2CO' : 2CB$,

$\therefore CO \cdot CO' = CB^2$.

Def. When the straight line AB and therefore also C is given points O and O' chosen so that

$$CO \cdot CO' = CB^2$$

are said to be in *involution*. Such points therefore divide AB harmonically.

73. **Def.** If a straight line AB be divided in *any* two points O and O', either both internal or both external, or one internal and the other external, and without any limitation as to the side of AB on which the external point or points should fall, the ratio of the ratios $AO : OB$ and $AO' : O'B$, or, as we shall express it,

the ratio $(AO : OB) : (AO' : O'B)$

is called the *an*harmonic ratio of the *range* $AOBO'$, which is always thus represented by $AOBO'$, whatever may be the *geometrical* order in which the letters O and O' appear with reference to A and B.

The *an*harmonic ratio of the range $AO'BO$ is

the ratio $(AO' : O'B) : (AO : OB)$,

and is therefore the reciprocal of the *an*harmonic ratio of the range $AOBO'$.

If the points O and O' are both internal or both external with regard to A and B, then the anharmonic ratio of the range $AOBO'$, viewed *algebraically*, is positive; when one point of division is external and the other internal it is negative. When therefore the straight line AB is *harmonically* divided in O and O', the anharmonic ratio of the range $AOBO'$ or $AO'BO$ would be *algebraically* represented by -1.

When the anharmonic ratios of the ranges $AOBO'$ and $AO'BO$ are equal, it is evident that AB is harmonically divided; for the only value of the anharmonic ratio, which can be the same as its reciprocal, except unity (which would imply that the points O and O' were coincident) is -1, and in this case, the line AB is harmonically divided.

If the points A, O, B, O' of a range $AOBO'$ be joined to a point E, external to AB, then the four straight lines thus formed are called a *pencil*, which is represented by $E(AOBO')$.

Prop. IV.

74. If the four straight lines EA, EO, EB, EO' forming a pencil $E(AOBO')$ be cut by *any* other straight line in the points a, o, b, o' respectively, then the anharmonic ratio of the range $aobo'$ will be the same as that of the range $AOBO'$, however the straight line $aobo'$ be drawn.

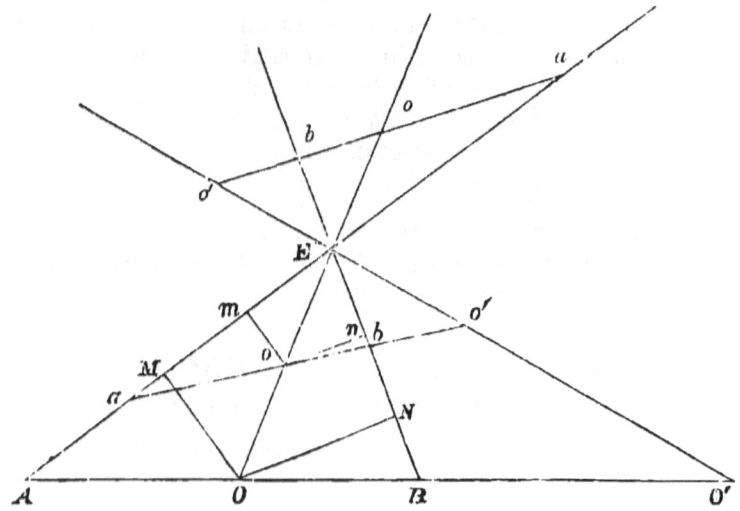

From the points O and o draw OM, om at right angles to EA, and ON, on at right angles to EB; then the ratio

$$(AO : OB) : (ao : ob) :: (\triangle AOE : BOE) : (\triangle aoE : boE)$$
$$:: (OM . AE : ON . EB) : (om . aE : on . Eb)$$
$$:: (AE : EB) : (aE : Eb).$$

So also for the point O'

$$(AO' : O'B) : (ao' : o'b) :: (AE : EB) : (aE : Eb)$$
$$\therefore (AO : OB) : (ao : ob) :: (AO' : O'B) : (ao' : o'b),$$
or $(AO : OB) : (AO' : O'B) :: (ao : ob) : (ao' : o'b).$

Prop. V.

The straight lines joining the intersections of the diagonals of a quadrilateral figure with the points of intersection of a pair of opposite sides are divided harmonically at the points where they meet the other two sides; also the sides of the quadrilateral are divided harmonically by these straight lines.

Let $ABCD$ be a quadrilateral, and let the diagonals AC, BD, intersect in E, and the pairs of opposite sides AD, BC, and AB, DC, in F and G respectively.

Let EF meet the sides CD, AB, in P, Q; and EG the sides AD, BC, in R, S.

Since the four straight lines EA, EQ, EB, EG, forming the pencil $E(AQBG)$ are intersected by the straight line $DPCG$ in the points C, P, D, G, the ranges $AQBG$, $CPDG$, have the same anharmonic ratio.

Again, since the four straight lines FA, FQ, FB, FG, forming the pencil $F(AQBG)$ are intersected by the straight line $DPCG$ in the points D, P, C, G, the ranges $AQBG$, $DPCG$, have the same anharmonic ratio.

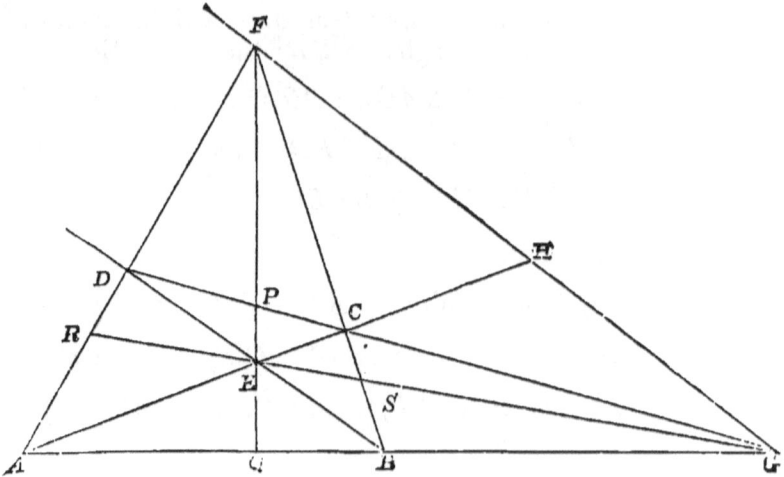

∴ the ranges $CPDG$, $DPCG$, have the same anharmonic ratio.

But the anharmonic ratio of the range

$CPDG$, viz. the ratio $(CP : PD) : (CG : GD)$,

is the reciprocal of the anharmonic ratio of the range,

$DPCG$, viz. the ratio $(DP : PC) : (DG : GC)$,

∴ the ranges $DPCG$ and $CPDG$ are harmonic (*Art.* 73).

Hence also the ranges $RESG$, $AQBG$, are harmonic; and exactly in the same way it may be proved that the ranges $FCSB$, $FPEQ$, $FDRA$, are harmonic.

Cor. If AC, BD, be produced to meet FG in H and K respectively; then

AC, BD, are divided harmonically in E, H, and E, K respectively; and

GF is divided harmonically in H, K.

N.B.—The point K will be on FG or GF produced, according as GH is less or greater than HF. If BE be equal to ED the point K is at infinity, and GF, BD are parallel, and GF is also bisected in H.

Prop. VI.

75. If from an external point O a pair of tangents OP, OP' be drawn to any conic, and a straight line OQQ' intersect the

curve in Q, Q' and PP' in O'; then QQ' is divided harmonically in O' and O.

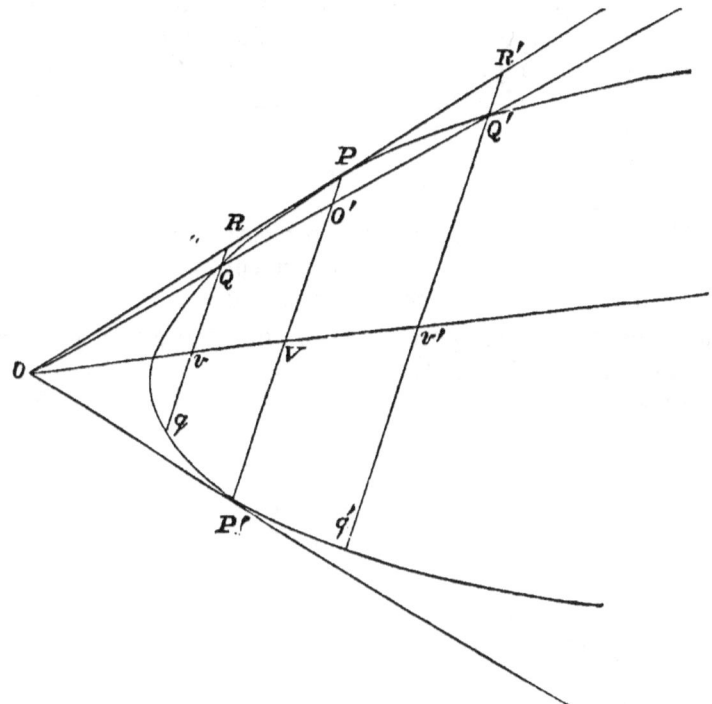

Through O draw OV bisecting PP' in V, and draw the double ordinates $RQvq$, $R'Q'v'q'$ parallel to PP' meeting OV in v and v'; then

Qq and $Q'q'$ are bisected in v and v'

Now
$$RP^2 : R'P^2 :: RQ \cdot Rq : R'Q' \cdot R'q'$$
$$:: Rv^2 - Qv^2 : R'v'^2 - Q'v'^2$$
$$:: Ov^2 : Ov'^2$$
$$:: OR^2 : OR'^2$$
$\therefore RP : R'P :: OR : OR'$
$\therefore QO' : O'Q' :: QO : OQ'$

$\therefore QQ'$ is divided harmonically in O, O'; and therefore also (see Prop. II.) OO' is divided harmonically in Q and Q'.

160   APPENDIX.

Cor. If the conic is a parabola, OV is drawn parallel to the axis. If an ellipse or hyperbola, OV is drawn through the centre. If the point O be the centre of the hyperbola, then, OP, OP' are asymptotes, and the line PP' is *at infinity*, and any chord QQ' through C is bisected at C, while the fourth point which with C harmonically divides QQ' is at an infinite distance.

Prop. VII.

76. The locus of the point of intersection of the tangents at the extremities of *any* chord drawn through a given point either within or without a conic is a straight line.

Let O be the given point. Through O draw any chord QOQ', and bisect QQ' in V.

First let the curve be a parabola.

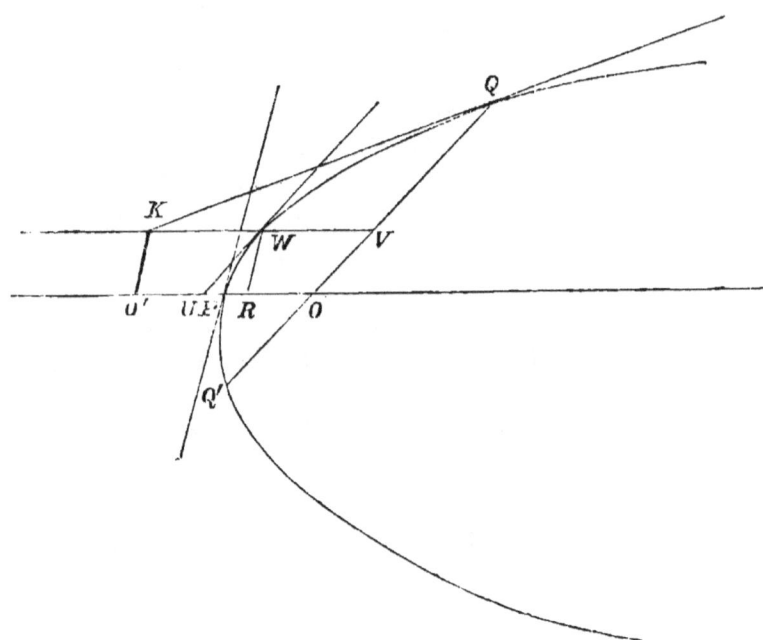

Through O and V draw OPO' and VWK parallel to the axis of the parabola meeting the curve in P and W.

Make WK equal to VW

and PO' equal to PO

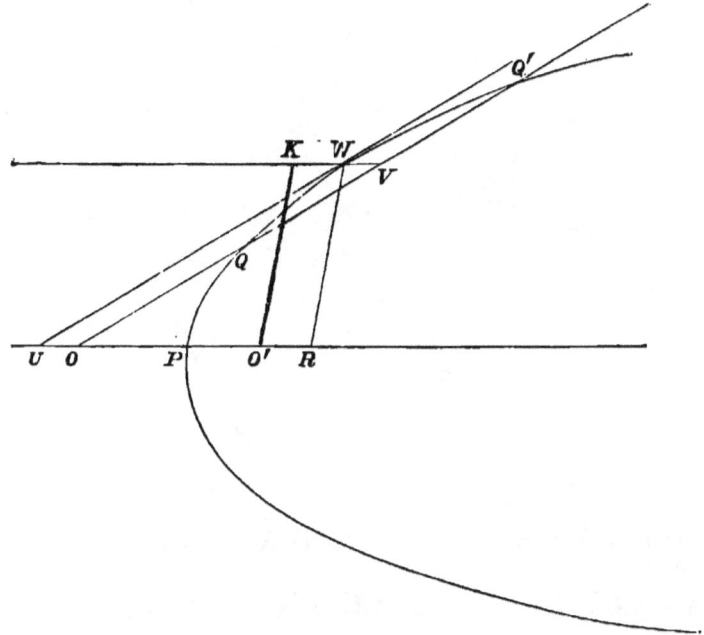

Then K is the point of intersection of the tangents at Q and Q'.

From W draw the tangent WU parallel to QQ' and the ordinate WR parallel to the tangent at P meeting OO' in U and R respectively.

Join KO'.

$$\begin{aligned} \text{Now } PO' &= PO \text{ by construction} \\ \text{and } PR &= PU \\ \therefore O'R &= OU \\ &= WV \\ &= KW \end{aligned}$$

$\therefore KO'$ is parallel to WR, and therefore to the tangent at P.

$\therefore$ the locus of K is the straight line drawn through O' parallel to the tangent at P.

Next let the curve be a central conic whose centre is C.

M

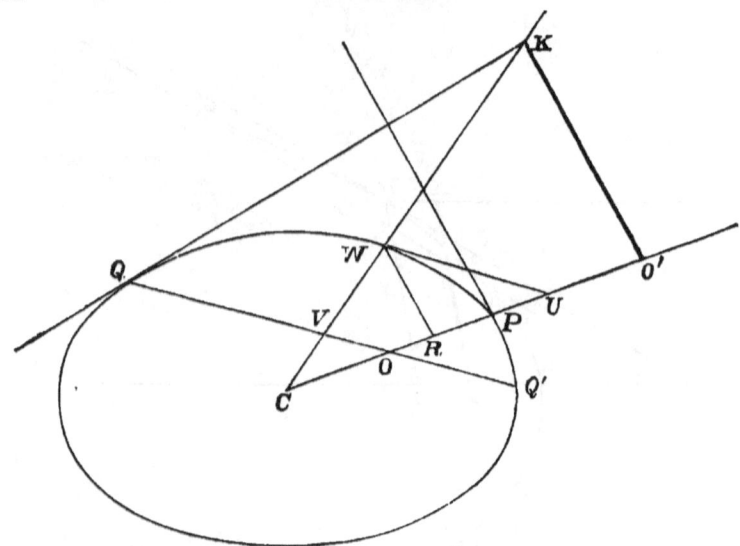

Through C draw $COPO'$, $CVWK$ meeting the conic in P and W.

Take CO', CK third proportionals respectively to CO, CP, and CV, CW, so that
$$CO \cdot CO' = CP^2 \text{ and } CV \cdot CK = CW^2.$$

Then K is the point of intersection of the tangents at Q and Q'.

Through W draw the tangent WU parallel to QQ', and the ordinate WR parallel to the tangent at P meeting CP in U and R respectively.

Join KO'.

$$\text{Now } CR \cdot CU = CP^2$$
$$= CO \cdot CO'$$
$$\therefore CR : CO' :: CO : CU$$
$$:: CV : CW$$
$$:: CW : CK \text{ by construction.}$$

$\therefore KO'$ is parallel to WR, and therefore to the tangent at P.

$\therefore$ the locus of K is the straight line drawn through O' parallel to the tangent at P.

N.B.—In the figure the conic has been drawn an ellipse and the point O has been taken on the inside, but it is quite as easy to draw the figure when the point O is on the outside of the curve, or when the conic is a hyperbola, and the point O either internal or external.

77. Def. The straight line KO' is called the *polar* of the point O.

If the point O is on the *concave* side of the conic, the polar is entirely *external*.

When the point O is on the *convex* side of the curve, the polar intersects the conic in two points, and is identical with the chord joining the points of contact of the pair of tangents drawn from O.

For if the chord OQQ' drawn through the external point O move round the point O until Q, Q' coincide and OQQ' becomes a tangent, the points K, Q, Q', will evidently all coincide with the point of contact of the tangent drawn from O. Hence the points of contact of the tangents drawn from O are on the *polar*, and the *polar* is identical with the chord joining the points of contact.

78. When the pole is given the mode of constructing geometrically the polar, or, when the polar is given, of finding the pole is evident at once from the above proposition, whether the curve be a parabola or a central conic.

Prop. VIII.

If the polar of O pass through o, then the polar of o passes through O.

The points O and o may be either both *external*, or one *internal* and the other *external*, but they cannot be both internal; for the polar of an *internal* point is wholly *external*.

(1) Let O be external. Then the polar of O is the chord joining the points of contact of tangents drawn from O; and since this chord passes through o, and O is the point of intersection of the tangents at the extremities of this chord, O is evidently on the polar of o.

(2) Let O be internal. Then since the point o is on the polar of O, the chord of contact of the pair of tangents drawn from o, *i.e.* the polar of o, will pass through the point O.

Cor. From this it is evident that the point of intersection of two polars to a conic is the pole of the line joining the two poles.

Prop. IX.

Any chord of a conic is divided harmonically by any point upon it, and the point where the polar of this point meets the chord.

Let QQ' be any chord of a conic, and O any point upon it.

(1) If the point O is *external* this proposition is already proved by Prop. VI.

(2) If the point O is on the concave side of the conic, let the polar of O intersect QQ' or QQ' produced in O'; then since the point O' is on the convex side of the curve, and is on the polar of O, the polar of O' *i.e.* the chord joining the points of contact of the pair of tangents drawn from O' must pass through O, and therefore as before QQ' is divided harmonically in O and O'.

79. If any number of points be on a straight line, their polars with respect to any conic will pass through the same point, viz., the pole of the straight line on which the points all lie; and conversely if any number of straight lines pass through a point, their poles all lie on the same straight line, viz., the polar of the given point. This is briefly expressed by saying that if any number of points are *collinear*, their polars are *confocal*, and conversely if any number of lines are *confocal*, their poles are *collinear*.

If a point instead of moving upon a straight line trace out a curve, the polars of the various points of the curve with respect to any conic (which in this case is called the *auxiliary* conic) will by their ultimate intersections form a new curve to which they are all tangents.

Thus if P and Q be two contiguous points on the original curve and p and q represent their polars (with respect to any auxiliary conic) the intersection of the straight lines p and q will *ultimately* be a point on the new curve; and since this point will be the pole of PQ which when P and Q are close together is *ultimately* a tangent to the original curve, it is evident that the polars of the several points of the new curve are tangents of the original curve, which can therefore be derived from the second curve in exactly the same manner as the second was derived from the first.

On account of this reciprocal property the locus of the ultimate intersections or the envelope of (*i.c.* the curve touched by) the polars of the various points of a given curve, is called the polar reciprocal of the proposed curve.

In *analytical* treatises it is shown that the polar reciprocal of any conic (with respect to an auxiliary conic) is also a conic.

APPENDIX.   165

In this brief notice we shall confine ourselves to shewing that any conic section may be produced by reciprocating, *i.e.* taking the polar reciprocal of a *circle* with respect to an *auxiliary* circle.

The advantage of having a circle for the *auxiliary* conic consists in the fact that the straight line drawn from any point to the centre of the circle is perpendicular to the polar of the point, and therefore also that the angle subtended at the centre of the circle by any two points is equal to one of the angles contained by the polars of the points.

In finding the polar reciprocal of a given curve, there are two ways in which we may proceed. We may either find the envelope of the polars of the different points of the original curve, or we may find the locus of the poles of the tangents at the different points of the original curve. The latter is the method which will be followed in the next article.

Prop. X.

The polar reciprocal of a circle, with respect to an auxiliary circle, will be a conic.

Let S be the centre of the auxiliary circle, and O the centre of the circle to be reciprocated.

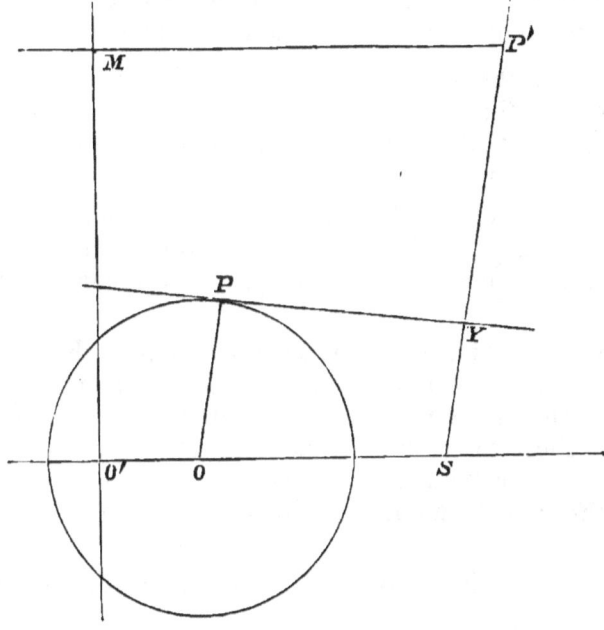

Join SO, and on SO produced, if necessary, take SO' a third proportional to SO and the radius of the auxiliary circle. Through O' draw $O'M$ at right angles to SO', then $O'M$ is the polar of O (Art. 78).

At any point P on the circle whose centre is O, draw the tangent PY; and from S draw SY perpendicular to PY.

On SY produced, if necessary, take a point P' so that
$$SY \cdot SP' = SO \cdot SO'$$
Then P' is the pole of PY.

Draw $P'M$ perpendicular to $O'M$; then since
$$SY \cdot SP' = SO \cdot SO'$$
$$\therefore SP' : SO' :: SO : SY$$

Now since the quadrilateral figures $SP'MO'$, $SOPY$ are equiangular and
$$SP' : SO' :: SO : SY$$
it is at once seen that these figures are also *similar*.
$$\therefore SP' : P'M :: SO : OP$$

$\therefore$ the locus of P' is a conic whose focus is S, directrix $O'M$ the polar of O, and eccentricity the ratio of $SO : OP$.

Cor. Since the eccentricity of the conic depends only upon the ratio of $SO : OP$, the polar reciprocal will be an ellipse, parabola, or hyperbola, according as S is inside, on, or outside the circle to be reciprocated.

The distance of the directrix from S can be made as large or small as we please by increasing or diminishing the radius of the auxiliary circle, without altering the eccentricity of the curve.

If the point P be such that the tangent pass through S, the point P' will be at infinity. If therefore tangents be drawn from S to the given circle, the lines drawn through S at right angles to these tangents will meet the curve at infinity, and will therefore be parallel to the asymptotes of the conic, which in this case will be an hyperbola since S is outside the given circle.

The focus S, the distance SO' of the directrix, and the

APPENDIX.

eccentricity being known, all the other elements of the conic can be at once determined.

80. By combining the harmonic properties of the complete quadrilateral with the harmonic property of the conic, viz., that any chord is divided harmonically by the pole and polar, many important theorems with regard to figures inscribed in and circumscribed about a conic may be deduced.

Prop. XI.

The triangle formed by the point of intersection of the diagonals of a quadrilateral figure inscribed is a conic, and the points of intersection of the opposite sides is self-conjugate, *i.e.* each angular point is the pole (with regard to the conic) of the opposite side.

Let $ABCD$ be a quadrilateral inscribed in a conic.

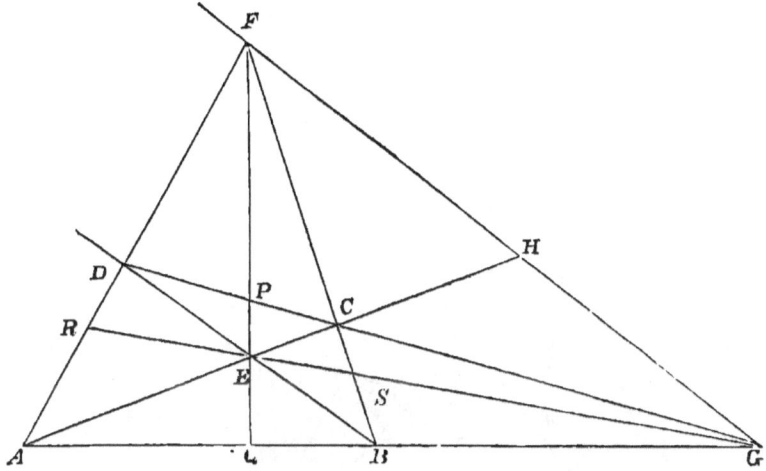

Now since the chord BC is divided harmonically in F and S, (*Prop.* V.)

∴ the polar of F passes through S.
So the polar of F passes through R.
∴ RS or EG is the polar of F.

168   APPENDIX.

So also EF is the polar of G.

Hence E the point of intersection of EF and EG is the polar of FG and $\therefore$ the triangle EFG is self-conjugate.

Prop. XII.

The intersection of the diagonals of a quadrilateral figure described about a conic is the pole with respect to the conic of the straight line joining the points of intersection of the pairs of opposite sides.

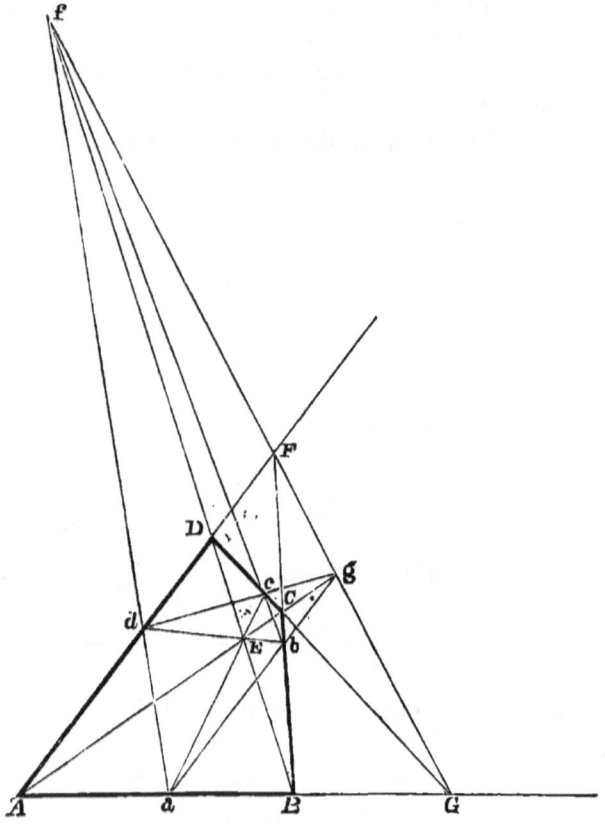

Let $ABCD$ be a quadrilateral figure described about a

conic; then the intersection E of the diagonals AC, BD shall be the pole of the straight line FG joining the points of intersection of AD, BC and AB, DC.

Let a, b, c, d be the points of contact of the sides of the quadrilateral with the conic.

Complete the quadrilateral $abcd$, and let e be the intersection of the diagonals ac, bd; and f and g those of the pairs of the opposite sides ad, bc and dc, ab respectively.

Now by the last proposition e is the pole of fg with respect to the conic which is inscribed in $ABCD$ and therefore described about $abcd$.

But since F is the pole of db and G is the pole of ac
$\therefore$ e is the pole of FG
$\therefore$ the straight lines fg and FG coincide.

Again since A is the pole of ad and C is the pole of bc
$\therefore$ f is the pole of AC
So also g is the pole of BD
$\therefore$ E is the pole of fg
$\therefore$ the points E and e coincide,
but e is the pole of FG
$\therefore$ E is the pole of FG.

Cor. 1. Since f is the pole of AC and also of eg or Eg, the straight line AC passes through g
So also BD passes through f.

Cor. 2. If ac pass through the point F then G is the pole of EF, and
E has been proved to be the pole of FG
$\therefore$ F is the pole of EG
$\therefore$ bd passes through the point G.

In this case, but in this only, the $\triangle EFG$ is self-conjugate.

LONDON:
R. CLAY, SONS, AND TAYLOR,
BREAD STREET HILL, E.C.

MESSRS. MACMILLAN AND CO.'S PUBLICATIONS.

SOLUTIONS TO PROBLEMS CONTAINED IN A GEO-
METRICAL TREATISE ON CONIC SECTIONS. By the Rev. W. H. DREW, M.A. of St. John's College, Cambridge, &c. Crown 8vo. 4s. 6d.

AN ELEMENTARY TREATISE ON CONIC SECTIONS.
By CHARLES SMITH, M.A., Fellow and Tutor of Sidney Sussex College, Cambridge. Crown 8vo. 7s. 6d.

By J. M. WILSON, M.A., late Fellow of St. John's College, Cambridge; and Head Master of Clifton College.

    ELEMENTARY GEOMETRY. BOOKS I. to V.—Containing the subjects of Euclid's First Six Books; following the Syllabus of Geometry prepared by the Geometrical Association. New Edition, enlarged. Extra fcap. 8vo. 4s. 6d.

    SOLID GEOMETRY AND CONIC SECTIONS. With Appendices on Transversals, and Harmonic Division, for the use of Schools. New Edition. Extra fcap. 8vo. 3s. 6d.

PLANE TRIGONOMETRY. GRADUATED EXERCISES
IN PLANE TRIGONOMETRY. Compiled and arranged by J. WILSON, M.A., Fellow of Christ's College, Cambridge, and S. R. WILSON, B.A., Fellow of Sidney Sussex College, Cambridge. Crown 8vo. 4s. 6d.

By I. TODHUNTER, M.A., F.R.S.

    EUCLID FOR COLLEGES AND SCHOOLS. 3s. 6d.—KEY, 6s. 6d.

    MENSURATION FOR BEGINNERS. 2s. 6d.

    ALGEBRA FOR BEGINNERS. With numerous Examples. 2s. 6d.—KEY, 6s. 6d.

    TRIGONOMETRY FOR BEGINNERS. 2s. 6d.—KEY, 8s. 6d.

    MECHANICS FOR BEGINNERS. 4s. 6d.—KEY, 6s. 6d.

    ALGEBRA FOR THE USE OF COLLEGES AND SCHOOLS. 7s. 6d.—KEY, 10s. 6d.

    THE THEORY OF EQUATIONS. 7s. 6d.

    PLANE TRIGONOMETRY. 5s.—KEY, 10s. 6d.

    SPHERICAL TRIGONOMETRY. 4s. 6d.

    CONIC SECTIONS. With Examples. 7s. 6d.

    THE DIFFERENTIAL CALCULUS. With Examples. 10s. 6d.

    THE INTEGRAL CALCULUS. 10s. 6d.

    EXAMPLES OF ANALYTICAL GEOMETRY OF THREE DIMENSIONS. 4s.

    ANALYTICAL STATICS. With Examples. 10s. 6d.

MACMILLAN AND CO., LONDON.

MESSRS. MACMILLAN & CO.'S PUBLICATIONS.

By BARNARD SMITH, M.A.
   ARITHMETIC AND ALGEBRA. New Edition. Crown 8vo. 10s. 6d.
   ARITHMETIC FOR THE USE OF SCHOOLS. New Edition. Crown 8vo. 4s. 6d.—KEY, Crown 8vo, 8s. 6d.
   EXERCISES IN ARITHMETIC. Crown 8vo. 2 Parts, 1s. each, or complete, 2s.—With Answers, 2s. 6d. Answers separately, 6d.
   SCHOOL CLASS-BOOK OF ARITHMETIC. 18mo. 3s. Or, sold separately, in Three Parts, 1s. each.—KEY, Three Parts, 2s. 6d. each.

ARITHMETIC IN THEORY AND PRACTICE. By J. BROOK-SMITH, M.A. New Edition, Revised. Crown 8vo. 4s. 6d.

ELEMENTARY TRIGONOMETRY. By Rev. J. B. LOCK, M.A., Fellow of Caius College, Cambridge; Assistant Master at Eton. With Diagrams. Globe 8vo. 4s. 6d.—PART II. HIGHER TRIGONOMETRY. [*In the Press.*

GEOMETRICAL EXERCISES FOR BEGINNERS. By SAMUEL CONSTABLE, ex-Sizar of Trinity College, Dublin; late Senior Mathematical Master, Corrig School, Kingstown. Crown 8vo. 3s. 6d.

NOTE-BOOK ON PRACTICAL SOLID OR DESCRIPTIVE GEOMETRY, CONTAINING PROBLEMS WITH HELP FOR SOLUTION. By J. H. EDGAR, M.A., Lecturer on Mechanical Drawing in the Royal School of Mines, London, and G. S. PRITCHARD. Fourth Edition, Enlarged. By ARTHUR G. MEEZE, Assistant Lecturer on Drawing, Royal School of Mines. Globe 8vo. 4s. 6d.

By S. PARKINSON, D.D., F.R.S.
   MECHANICS, AN ELEMENTARY TREATISE ON. With Examples. 9s. 6d.
   A TREATISE ON OPTICS. Crown 8vo. 10s. 6d.

By Prof. G. BOOLE, F.R.S.
   DIFFERENTIAL EQUATIONS. Crown 8vo. 14s. Supplementary Vol. 8s. 6d.
   CALCULUS OF FINITE DIFFERENCES. Crown 8vo. 10s. 6d.

By E. J. ROUTH, M.A., LL.D., F.R.S.
   A TREATISE ON THE DYNAMICS OF A SYSTEM OF RIGID BODIES, THE FIRST PART OF. With numerous Examples. Fourth Edition, Revised and Enlarged. 8vo. 14s. [*Part II. in the Press.*
   STABILITY OF A GIVEN STATE OF MOTION, PARTICULARLY STEADY MOTION. Adams Prize Essay for 1877. 8vo. 8s. 6d.

By W. K. CLIFFORD, F.R.S.
   ELEMENTS OF DYNAMICS. An Introduction to the Study of Motion and Rest in Solid and Fluid Bodies. Part I. Kinematic. Crown 8vo. 7s. 6d.
   MATHEMATICAL PAPERS. Edited by R. TUCKER. With an Introduction by H. J. STEPHEN SMITH, M.A., LL.D., F.R.S., &c. Demy 8vo. 30s.

MACMILLAN AND CO., LONDON.

A Catalogue

OF WORKS ON

Mathematics, Science,

AND

History and Geography.

PUBLISHED BY

Macmillan & Co.,

BEDFORD STREET, STRAND, LONDON.

CONTENTS.

PAGE.

MATHEMATICS—
- Arithmetic 3
- Algebra 5
- Euclid and Elementary Geometry . . 5
- Mensuration 6
- Higher Mathematics 7

SCIENCE—
- Natural Philosophy 14
- Astronomy 19
- Chemistry 19
- Biology 21
- Medicine 25
- Anthropology 25
- Physical Geography and Geology . . . 25
- Agriculture 26
- Political Economy 27
- Mental and Moral Philosophy 27

HISTORY AND GEOGRAPHY 28

29 AND 30, BEDFORD STREET, COVENT GARDEN,
LONDON, W.C., *April* 1884.

MATHEMATICS.

(1) Arithmetic, (2) Algebra, (3) Euclid and Elementary Geometry, (4) Mensuration, (5) Higher Mathematics.

ARITHMETIC.

Aldis.—THE GIANT ARITHMOS. A most Elementary Arithmetic for Children. By MARY STEADMAN ALDIS. With Illustrations. Globe 8vo. 2*s*. 6*d*.

Brook-Smith (J.).—ARITHMETIC IN THEORY AND PRACTICE. By J. BROOK-SMITH, M.A., LL.B., St. John's College, Cambridge; Barrister-at-Law; one of the Masters of Cheltenham College. New Edition, revised. Crown 8vo. 4*s*. 6*d*.

Candler.—HELP TO ARITHMETIC. Designed for the use of Schools. By H. CANDLER, M.A., Mathematical Master of Uppingham School. Extra fcap. 8vo. 2*s*. 6*d*.

Dalton.—RULES AND EXAMPLES IN ARITHMETIC. By the Rev. T. DALTON, M.A., Assistant-Master of Eton College. New Edition. 18mo. 2*s*. 6*d*.
[*Answers to the Examples are appended.*

Pedley.—EXERCISES IN ARITHMETIC for the Use of Schools. Containing more than 7,000 original Examples. By S. PEDLEY, late of Tamworth Grammar School. Crown 8vo. 5*s*.

Smith.—Works by the Rev. BARNARD SMITH, M.A., late Rector of Glaston, Rutland, and Fellow and Senior Bursar of S. Peter's College, Cambridge.
ARITHMETIC AND ALGEBRA, in their Principles and Application; with numerous systematically arranged Examples taken from the Cambridge Examination Papers, with especial reference to the Ordinary Examination for the B.A. Degree. New Edition, carefully Revised. Crown 8vo. 10*s*. 6*d*.
ARITHMETIC FOR SCHOOLS. New Edition. Crown 8vo. 4*s*. 6*d*.
A KEY TO THE ARITHMETIC FOR SCHOOLS. New Edition. Crown 8vo. 8*s*. 6*d*.

Smith.—Works by the Rev. BARNARD SMITH, M.A. *(continued)*—

EXERCISES IN ARITHMETIC. Crown 8vo, limp cloth, 2s. With Answers, 2s. 6d.

Answers separately, 6d.

SCHOOL CLASS-BOOK OF ARITHMETIC. 18mo, cloth. 3s. Or sold separately, in Three Parts, 1s. each.

KEYS TO SCHOOL CLASS-BOOK OF ARITHMETIC. Parts I., II., and III., 2s. 6d. each.

SHILLING BOOK OF ARITHMETIC FOR NATIONAL AND ELEMENTARY SCHOOLS. 18mo, cloth. Or separately, Part I. 2d.; Part II. 3d.; Part III. 7d. Answers. 6d.

THE SAME, with Answers complete. 18mo, cloth. 1s. 6d.

KEY TO SHILLING BOOK OF ARITHMETIC. 18mo. 4s. 6d.

EXAMINATION PAPERS IN ARITHMETIC. 18mo. 1s. 6d. The same, with Answers, 18mo, 2s. Answers, 6d.

KEY TO EXAMINATION PAPERS IN ARITHMETIC. 18mo. 4s. 6d.

THE METRIC SYSTEM OF ARITHMETIC, ITS PRINCIPLES AND APPLICATIONS, with numerous Examples, written expressly for Standard V. in National Schools. New Edition. 18mo, cloth, sewed. 3d.

A CHART OF THE METRIC SYSTEM, on a Sheet, size 42 in. by 34 in. on Roller, mounted and varnished. New Edition. Price 3s. 6d.

Also a Small Chart on a Card, price 1d.

EASY LESSONS IN ARITHMETIC, combining Exercises in Reading, Writing, Spelling, and Dictation. Part I. for Standard I. in National Schools. Crown 8vo. 9d.

EXAMINATION CARDS IN ARITHMETIC. (Dedicated to Lord Sandon.) With Answers and Hints.

Standards I. and II. in box, 1s. Standards III., IV., and V., in boxes, 1s. each. Standard VI. in Two Parts, in boxes, 1s. each.

A and B papers, of nearly the same difficulty, are given so as to prevent copying, and the colours of the A and B papers differ in each Standard, and from those of every other Standard, so that a master or mistress can see at a glance whether the children have the proper papers.

ALGEBRA.

Dalton.—RULES AND EXAMPLES IN ALGEBRA. By the Rev. T. DALTON, M.A., Assistant-Master of Eton College. Part I. New Edition. 18mo. 2s. Part II. 18mo. 2s. 6d.

Jones and Cheyne.—ALGEBRAICAL EXERCISES. Progressively Arranged. By the Rev. C. A. JONES, M.A., and C. H. CHEYNE, M.A., F.R.A.S., Mathematical Masters of Westminster School. New Edition. 18mo. 2s. 6d.

Smith.—ARITHMETIC AND ALGEBRA, in their Principles and Application; with numerous systematically arranged Examples taken from the Cambridge Examination Papers, with especial reference to the Ordinary Examination for the B.A. Degree. By the Rev. BARNARD SMITH, M.A., late Rector of Glaston, Rutland, and Fellow and Senior Bursar of St. Peter's College, Cambridge. New Edition, carefully Revised. Crown 8vo. 10s. 6d.

Todhunter.—Works by I. TODHUNTER, M.A., F.R.S., D.Sc., late of St. John's College, Cambridge.

"Mr. Todhunter is chiefly known to Students of Mathematics as the author of a series of admirable mathematical text-books, which possess the rare qualities of being clear in style and absolutely free from mistakes, typographical or other."—SATURDAY REVIEW.

ALGEBRA FOR BEGINNERS. With numerous Examples. New Edition. 18mo. 2s. 6d.

KEY TO ALGEBRA FOR BEGINNERS. Crown 8vo. 6s. 6d.

ALGEBRA. For the Use of Colleges and Schools. New Edition. Crown 8vo. 7s. 6d.

KEY TO ALGEBRA FOR THE USE OF COLLEGES AND SCHOOLS. Crown 8vo. 10s. 6d.

EUCLID & ELEMENTARY GEOMETRY.

Constable.—GEOMETRICAL EXERCISES FOR BEGINNERS. By SAMUEL CONSTABLE. Crown 8vo. 3s. 6d.

Cuthbertson.—EUCLIDIAN GEOMETRY. By FRANCIS CUTHBERTSON, M.A., LL.D., Head Mathematical Master of the City of London School. Extra fcap. 8vo. 4s. 6d.

Dodgson.—EUCLID. BOOKS I. AND II. Edited by CHARLES L. DODGSON, M.A., Student and late Mathematical Lecturer of Christ Church, Oxford. Second Edition, with words substituted for the Algebraical Symbols used in the First Edition. Crown 8vo. 2s.

⁎ The text of this Edition has been ascertained, by counting the words, to be *less than five-sevenths* of that contained in the ordinary editions.

Kitchener.—A GEOMETRICAL NOTE-BOOK, containing Easy Problems in Geometrical Drawing preparatory to the Study of Geometry. For the use of Schools. By F. E. KITCHENER, M.A., Mathematical Master at Rugby. New Edition. 4to. 2s.

Mault.—NATURAL GEOMETRY: an Introduction to the Logical Study of Mathematics. For Schools and Technical Classes. With Explanatory Models, based upon the Tachymetrical works of Ed. Lagout. By A. MAULT. 18mo. 1s. Models to Illustrate the above, in Box, 12s. 6d.

Syllabus of Plane Geometry (corresponding to Euclid, Books I.—VI.). Prepared by the Association for the Improvement of Geometrical Teaching. New Edition. Crown 8vo. 1s.

Todhunter.—THE ELEMENTS OF EUCLID. For the Use of Colleges and Schools. By I. TODHUNTER, M.A., F.R.S., D.Sc., of St. John's College, Cambridge. New Edition. 18mo. 3s. 6d.

KEY TO EXERCISES IN EUCLID. Crown 8vo. 6s. 6d.

Wilson (J. M.).—ELEMENTARY GEOMETRY. BOOKS I.—V. Containing the Subjects of Euclid's first Six Books. Following the Syllabus of the Geometrical Association. By the Rev. J. M. WILSON, M.A., Head Master of Clifton College. New Edition. Extra fcap. 8vo. 4s. 6d.

MENSURATION.

Tebay.—ELEMENTARY MENSURATION FOR SCHOOLS. With numerous examples. By SEPTIMUS TEBAY, B.A., Head Master of Queen Elizabeth's Grammar School, Rivington. Extra fcap. 8vo. 3s. 6d.

Todhunter.—MENSURATION FOR BEGINNERS. By I. TODHUNTER, M.A., F.R.S., D.Sc., late of St. John's College, Cambridge. With Examples. New Edition. 18mo. 2s. 6d.

HIGHER MATHEMATICS.

Airy.—Works by Sir G. B. AIRY, K.C.B., formerly Astronomer-Royal:—

ELEMENTARY TREATISE ON PARTIAL DIFFERENTIAL EQUATIONS. Designed for the Use of Students in the Universities. With Diagrams. Second Edition. Crown 8vo. 5s. 6d.

ON THE ALGEBRAICAL AND NUMERICAL THEORY OF ERRORS OF OBSERVATIONS AND THE COMBINATION OF OBSERVATIONS. Second Edition, revised. Crown 8vo. 6s. 6d.

Alexander (T.).—ELEMENTARY APPLIED MECHANICS. Being the simpler and more practical Cases of Stress and Strain wrought out individually from first principles by means of Elementary Mathematics. By T. ALEXANDER, C.E., Professor of Civil Engineering in the Imperial College of Engineering, Tokei, Japan. Crown 8vo. Part I. 4s. 6d.

Alexander and Thomson.—ELEMENTARY APPLIED MECHANICS. By THOMAS ALEXANDER, C.E., Professor of Engineering in the Imperial College of Engineering, Tokei, Japan; and ARTHUR WATSON THOMSON, C.E., B.Sc., Professor of Engineering at the Royal College, Cirencester. Part II. TRANSVERSE STRESS. Crown 8vo. 10s. 6d.

Bayma.—THE ELEMENTS OF MOLECULAR MECHANICS. By JOSEPH BAYMA, S.J., Professor of Philosophy, Stonyhurst College. Demy 8vo. 10s. 6d.

Beasley.—AN ELEMENTARY TREATISE ON PLANE TRIGONOMETRY. With Examples. By R. D. BEASLEY, M.A. Eighth Edition, revised and enlarged. Crown 8vo. 3s. 6d.

Blackburn (Hugh).—ELEMENTS OF PLANE TRIGONOMETRY, for the use of the Junior Class in Mathematics in the University of Glasgow. By HUGH BLACKBURN, M.A., late Professor of Mathematics in the University of Glasgow. Globe 8vo. 1s. 6d.

Boole.—Works by G. BOOLE, D.C.L., F.R.S., late Professor of Mathematics in the Queen's University, Ireland.

A TREATISE ON DIFFERENTIAL EQUATIONS. Third and Revised Edition. Edited by I. TODHUNTER. Crown 8vo. 14s.

Boole.—Works by G. BOOLE, D.C.L., &c. (*continued*)—
A TREATISE ON DIFFERENTIAL EQUATIONS. Supplementary Volume. Edited by I. TODHUNTER. Crown 8vo. 8s. 6d.
THE CALCULUS OF FINITE DIFFERENCES. Third Edition, revised by J. F. MOULTON. Crown 8vo. 10s. 6d.

Cambridge Senate-House Problems and Riders, with Solutions:—
1875—PROBLEMS AND RIDERS. By A. G. GREENHILL, M.A. Crown 8vo. 8s. 6d.
1878—SOLUTIONS OF SENATE-HOUSE PROBLEMS. By the Mathematical Moderators and Examiners. Edited by J. W. L. GLAISHER, M.A., Fellow of Trinity College, Cambridge. 12s.

Cheyne.—AN ELEMENTARY TREATISE ON THE PLANETARY THEORY. By C. H. H. CHEYNE, M.A., F.R.A.S. With a Collection of Problems. Third Edition. Edited by Rev. A. FREEMAN, M.A., F.R.A.S. Crown 8vo. 7s. 6d.

Christie.—A COLLECTION OF ELEMENTARY TEST-QUESTIONS IN PURE AND MIXED MATHEMATICS; with Answers and Appendices on Synthetic Division, and on the Solution of Numerical Equations by Horner's Method. By JAMES R. CHRISTIE, F.R.S., Royal Military Academy, Woolwich. Crown 8vo. 8s. 6d.

Clausius.—MECHANICAL THEORY OF HEAT. By R. CLAUSIUS. Translated by WALTER R. BROWNE, M.A., late Fellow of Trinity College, Cambridge. Crown 8vo. 10s. 6d.

Clifford.—THE ELEMENTS OF DYNAMIC. An Introduction to the Study of Motion and Rest in Solid and Fluid Bodies. By W. K. CLIFFORD, F.R.S., late Professor of Applied Mathematics and Mechanics at University College, London. Part I.—KINEMATIC. Crown 8vo. 7s. 6d.

Cotterill.—A TREATISE ON APPLIED MECHANICS. By JAMES COTTERILL, M.A., F.R.S., Professor of Applied Mechanics at the Royal Naval College, Greenwich. With Illustrations. 8vo.
[*In the press.*

Day. — PROPERTIES OF CONIC SECTIONS PROVED GEOMETRICALLY. Part I. THE ELLIPSE. With Problems. By the Rev. H. G. DAY, M.A. 8vo. 3s. 6d.

Day (R. E.)—ELECTRIC LIGHT ARITHMETIC. By R. E. DAY, M.A., Evening Lecturer in Experimental Physics at King's College, London. Pott 8vo. 2s.

MATHEMATICS.

Drew.—GEOMETRICAL TREATISE ON CONIC SECTIONS. By W. H. DREW, M.A., St. John's College, Cambridge. New Edition, enlarged. Crown 8vo. 5s.
SOLUTIONS TO THE PROBLEMS IN DREW'S CONIC SECTIONS. Crown 8vo. 4s. 6d.

Dyer.—EXERCISES IN ANALYTICAL GEOMETRY. Compiled and arranged by J. M. DYER, M.A., Senior Mathematical Master in the Classical Department of Cheltenham College. With Illustrations. Crown 8vo. 4s. 6d.

Edgar (J. H.) and Pritchard (G. S.).—NOTE-BOOK ON PRACTICAL SOLID OR DESCRIPTIVE GEOMETRY. Containing Problems with help for Solutions. By J. H. EDGAR, M.A., Lecturer on Mechanical Drawing at the Royal School of Mines, and G. S. PRITCHARD. Fourth Edition, revised by ARTHUR MEEZE. Globe 8vo. 4s. 6d.

Ferrers.—Works by the Rev. N. M. FERRERS, M.A., Fellow and Master of Gonville and Caius College, Cambridge.
AN ELEMENTARY TREATISE ON TRILINEAR CO-ORDINATES, the Method of Reciprocal Polars, and the Theory of Projectors. New Edition, revised. Crown 8vo. 6s. 6d.
AN ELEMENTARY TREATISE ON SPHERICAL HARMONICS, AND SUBJECTS CONNECTED WITH THEM. Crown 8vo. 7s. 6d.

Frost.—Works by PERCIVAL FROST, M.A., D.Sc., formerly Fellow of St. John's College, Cambridge; Mathematical Lecturer at King's College.
AN ELEMENTARY TREATISE ON CURVE TRACING. By PERCIVAL FROST, M.A. 8vo. 12s.
SOLID GEOMETRY. A New Edition, revised and enlarged, of the Treatise by FROST and WOLSTENHOLME. In 2 Vols. Vol. I. 8vo. 16s.

Hemming.—AN ELEMENTARY TREATISE ON THE DIFFERENTIAL AND INTEGRAL CALCULUS, for the Use of Colleges and Schools. By G. W. HEMMING, M.A., Fellow of St. John's College, Cambridge. Second Edition, with Corrections and Additions. 8vo. 9s.

Jackson.—GEOMETRICAL CONIC SECTIONS. An Elementary Treatise in which the Conic Sections are defined as the Plane Sections of a Cone, and treated by the Method of Projection. By J. STUART JACKSON, M.A., late Fellow of Gonville and Caius College, Cambridge. Crown 8vo. 4s. 6d.

Jellet (John H.).—A TREATISE ON THE THEORY OF FRICTION. By JOHN H. JELLET, B.D., Provost of Trinity College, Dublin; President of the Royal Irish Academy. 8vo. 8s. 6d.

Johnson.—INTEGRAL CALCULUS, an Elementary Treatise on the; Founded on the Method of Rates or Fluxions. By WILLIAM WOOLSEY JOHNSON, Professor of Mathematics at the United States Naval Academy, Annopolis, Maryland. Demy 8vo. 8s.

Kelland and Tait.—INTRODUCTION TO QUATERNIONS, with numerous examples. By P. KELLAND, M.A., F.R.S., and P. G. TAIT, M.A., Professors in the Department of Mathematics in the University of Edinburgh. Second Edition. Crown 8vo. 7s. 6d.

Kempe.— OW TO DRAW A STRAIGHT LINE: a Lecture on Linkages. By A. B. KEMPE. With Illustrations. Crown 8vo. 1s. 6d. (*Nature Series.*)

Lock.—ELEMENTARY TRIGONOMETRY. By Rev. J. B. LOCK, M.A., Senior Fellow, Assistant Tutor and Lecturer in Mathematics, of Gonville and Caius College, Cambridge; late Assistant-Master at Eton. Globe 8vo. 4s. 6d.
HIGHER TRIGONOMETRY. By the same Author. Globe 8vo. 3s. 6d.
Both Parts complete in One Volume. Globe 8vo. 7s. 6d.

Lupton.—ELEMENTARY CHEMICAL ARITHMETIC. With 1,100 Problems. By SYDNEY LUPTON, M.A., Assistant-Master in Harrow School. Globe 8vo. 5s.

Merriman.—ELEMENTS OF THE METHOD OF LEAST SQUARE. By MANSFIELD MERRIMAN, Ph.D., Professor of Civil and Mechanical Engineering, Lehigh University, Bethlehem, Penn. Crown 8vo. 7s. 6d.

Morgan.—A COLLECTION OF PROBLEMS AND EXAMPLES IN MATHEMATICS. With Answers. By H. A. MORGAN, M.A., Sadlerian and Mathematical Lecturer of Jesus College, Cambridge. Crown 8vo. 6s. 6d.

Millar.—ELEMENTS OF DESCRIPTIVE GEOMETRY. By J. B. MILLAR, C.E., Assistant Lecturer in Engineering in Owens College, Manchester. Crown 8vo. 6s.

Muir.—A TREATISE ON THE THEORY OF DETERMINANTS. With graduated sets of Examples. For use in Colleges and Schools. By THOS. MUIR, M.A., F.R.S.E., Mathematical Master in the High School of Glasgow. Crown 8vo. 7s. 6d.

Parkinson.—AN ELEMENTARY TREATISE ON MECHANICS. For the Use of the Junior Classes at the University and the Higher Classes in Schools. By S. PARKINSON, D.D., F.R.S., Tutor and Prælector of St. John's College, Cambridge. With a Collection of Examples. Sixth Edition, revised. Crown 8vo. 9s. 6d.

Phear.—ELEMENTARY HYDROSTATICS. With Numerous Examples. By J. B. PHEAR, M.A., Fellow and late Assistant Tutor of Clare College, Cambridge. New Edition. Crown 8vo. 5s. 6d.

Pirie.—LESSONS ON RIGID DYNAMICS. By the Rev. G. PIRIE, M.A., late Fellow and Tutor of Queen's College, Cambridge; Professor of Mathematics in the University of Aberdeen. Crown 8vo. 6s.

Rrice and Johnson.—DIFFERENTIAL CALCULUS, an Elementary Treatise on the; Founded on the Method of Rates or Fluxions. By JOHN MINOT PRICE, Professor of Mathematics in the United States Navy, and WILLIAM WOOLSEY JOHNSON, Professor of Mathematics at the United States Naval Academy. Third Edition, Revised and Corrected. Demy 8vo. 16s. Abridged Edition, 8s.

Puckle.—AN ELEMENTARY TREATISE ON CONIC SECTIONS AND ALGEBRAIC GEOMETRY. With Numerous Examples and Hints for their Solution; especially designed for the Use of Beginners. By G. H. PUCKLE, M.A. New Edition, revised and enlarged. Crown 8vo. 7s. 6d.

Rawlinson.—ELEMENTARY STATICS. By the Rev. GEORGE RAWLINSON, M.A. Edited by the Rev. EDWARD STURGES, M.A. Crown 8vo. 4s. 6d.

Reynolds.—MODERN METHODS IN ELEMENTARY GEOMETRY. By E. M. REYNOLDS, M.A., Mathematical Master in Clifton College. Crown 8vo. 3s. 6d.

Reuleaux.—THE KINEMATICS OF MACHINERY. Outlines of a Theory of Machines. By Professor F. REULEAUX. Translated and Edited by Professor A. B. W. KENNEDY, C.E. With 450 Illustrations. Medium 8vo. 21s.

Robinson.—TREATISE ON MARINE SURVEYING. Prepared for the use of younger Naval Officers. With Questions for Examinations and Exercises principally from the Papers of the

Robinson—*(continued)*—
>Royal Naval College. With the results. By Rev. JOHN I. ROBINSON, Chaplain and Instructor in the Royal Naval College, Greenwich. With Illustrations. Crown 8vo. 7s. 6d.
>CONTENTS.—Symbols used in Charts and Surveying—The Construction and Use of Scales—Laying off Angles—Fixing Positions by Angles — Charts and Chart-Drawing—Instruments and Observing — Base Lines—Triangulation—Levelling—Tides and Tidal Observations—Soundings—Chronometers—Meridian Distances—Method of Plotting a Survey—Miscellaneous Exercises—Index.

Routh.—Works by EDWARD JOHN ROUTH, M.A., F.R.S., D.Sc., late Fellow and Assistant Tutor at St. Peter's College, Cambridge; Examiner in the University of London.
>A TREATISE ON THE DYNAMICS OF THE SYSTEM OF RIGID BODIES. With numerous Examples. Fourth and enlarged Edition. Two Vols. Vol. I.—Elementary Parts. 8vo. 14s. Vol. II.—The Higher Parts. 8vo. *[In the press.*
>STABILITY OF A GIVEN STATE OF MOTION, PARTICULARLY STEADY MOTION. Adams' Prize Essay for 1877. 8vo. 8s. 6d.

Smith (C.).—CONIC SECTIONS. By CHARLES SMITH, M.A., Fellow and Tutor of Sidney Sussex College, Cambridge. Second Edition. Crown 8vo. 7s. 6d.

Snowball.—THE ELEMENTS OF PLANE AND SPHERICAL TRIGONOMETRY; with the Construction and Use of Tables of Logarithms. By J. C. SNOWBALL, M.A. New Edition. Crown 8vo. 7s. 6d.

Tait and Steele.—A TREATISE ON DYNAMICS OF A PARTICLE. With numerous Examples. By Professor TAIT and Mr. STEELE. Fourth Edition, revised. Crown 8vo. 12s.

Thomson.—A TREATISE ON THE MOTION OF VORTEX RINGS. An Essay to which the Adams Prize was adjudged in 1882 in the University of Cambridge. By J. J. THOMSON, Fellow and Assistant Lecturer of Trinity College, Cambridge. With Diagrams. 8vo. 6s.

Todhunter.—Works by I. TODHUNTER, M.A., F.R.S., D.Sc., late of St. John's College, Cambridge.
>"Mr. Todhunter is chiefly known to students of Mathematics as the author of a series of admirable mathematical text-books, which possess the rare qualities of being clear in style and absolutely free from mistakes, typographical and other."—SATURDAY REVIEW.
>TRIGONOMETRY FOR BEGINNERS. With numerous Examples. New Edition. 18mo. 2s. 6d.
>KEY TO TRIGONOMETRY FOR BEGINNERS. Crown 8vo. 8s. 6d.

MATHEMATICS. 13

Todhunter.—Works by I. TODHUNTER, M.A., &c. (*continued*)—
MECHANICS FOR BEGINNERS. With numerous Examples. New Edition. 18mo. 4s. 6d.
KEY TO MECHANICS FOR BEGINNERS. Crown 8vo. 6s. 6d.
AN ELEMENTARY TREATISE ON THE THEORY OF EQUATIONS. New Edition, revised. Crown 8vo. 7s. 6d.
PLANE TRIGONOMETRY. For Schools and Colleges. New Edition. Crown 8vo. 5s.
KEY TO PLANE TRIGONOMETRY. Crown 8vo. 10s. 6d.
A TREATISE ON SPHERICAL TRIGONOMETRY. New Edition, enlarged. Crown 8vo. 4s. 6d.
PLANE CO-ORDINATE GEOMETRY, as applied to the Straight Line and the Conic Sections. With numerous Examples. New Edition, revised and enlarged. Crown 8vo. 7s. 6d.
A TREATISE ON THE DIFFERENTIAL CALCULUS. With numerous Examples. New Edition. Crown 8vo. 10s. 6d.
A TREATISE ON THE INTEGRAL CALCULUS AND ITS APPLICATIONS. With numerous Examples. New Edition, revised and enlarged. Crown 8vo. 10s. 6d.
EXAMPLES OF ANALYTICAL GEOMETRY OF THREE DIMENSIONS. New Edition, revised. Crown 8vo. 4s.
A TREATISE ON ANALYTICAL STATICS. With numerous Examples. New Edition, revised and enlarged. Crown 8vo. 10s. 6d.
A HISTORY OF THE MATHEMATICAL THEORY OF PROBABILITY, from the time of Pascal to that of Laplace. 8vo. 18s.
RESEARCHES IN THE CALCULUS OF VARIATIONS, principally on the Theory of Discontinuous Solutions: an Essay to which the Adams' Prize was awarded in the University of Cambridge in 1871. 8vo. 6s.
A HISTORY OF THE MATHEMATICAL THEORIES OF ATTRACTION, AND THE FIGURE OF THE EARTH, from the time of Newton to that of Laplace. 2 vols. 8vo. 24s.
AN ELEMENTARY TREATISE ON LAPLACE'S, LAME'S, AND BESSEL'S FUNCTIONS. Crown 8vo. 10s. 6d.

Wilson (J. M.).—SOLID GEOMETRY AND CONIC SECTIONS. With Appendices on Transversals and Harmonic Division. For the Use of Schools. By Rev. J. M. WILSON, M.A. Head Master of Clifton College. New Edition. Extra fcap. 8vo. 3s. 6d.

Wilson.—GRADUATED EXERCISES IN PLANE TRIGONOMETRY. Compiled and arranged by J. WILSON, M.A., and S. R. WILSON, B.A. Crown 8vo. 4s. 6d.

"The exercises seem beautifully graduated and adapted to lead a student on most gently and pleasantly."—E. J. ROUTH, F.R.S., St. Peter's College, Cambridge.

(See also *Elementary Geometry*.)

Wilson (W. P.).—A TREATISE ON DYNAMICS. By W. P. WILSON, M.A., Fellow of St. John's College, Cambridge, and Professor of Mathematics in Queen's College, Belfast. 8vo. 9s. 6d.

Woolwich Mathematical Papers, for Admission into the Royal Military Academy, Woolwich, 1880—1883 inclusive. Crown 8vo. 3s. 6d.

Wolstenholme.—MATHEMATICAL PROBLEMS, on Subjects included in the First and Second Divisions of the Schedule of subjects for the Cambridge Mathematical Tripos Examination. Devised and arranged by JOSEPH WOLSTENHOLME, D.Sc., late Fellow of Christ's College, sometime Fellow of St. John's College, and Professor of Mathematics in the Royal Indian Engineering College. New Edition, greatly enlarged. 8vo. 18s.

EXAMPLES FOR PRACTICE IN THE USE OF SEVEN-FIGURE LOGARITHMS. By the same Author. [*In preparation.*

SCIENCE.

(1) Natural Philosophy, (2) Astronomy, (3) Chemistry, (4) Biology, (5) Medicine, (6) Anthropology, (7) Physical Geography and Geology, (8) Agriculture, (9) Political Economy, (10) Mental and Moral Philosophy.

NATURAL PHILOSOPHY.

Airy.—Works by Sir G. B. AIRY, K.C.B., formerly Astronomer-Royal :—

UNDULATORY THEORY OF OPTICS. Designed for the Use of Students in the University. New Edition. Crown 8vo. 6s. 6d.

ON SOUND AND ATMOSPHERIC VIBRATIONS. With the Mathematical Elements of Music. Designed for the Use of Students in the University. Second Edition, revised and enlarged. Crown 8vo. 9s.

A TREATISE ON MAGNETISM. Designed for the Use of Students in the University. Crown 8vo. 9s. 6d.

Airy (Osmond).—A TREATISE ON GEOMETRICAL OPTICS. Adapted for the Use of the Higher Classes in Schools. By OSMUND AIRY, B.A., one of the Mathematical Masters in Wellington College. Extra fcap. 8vo. 3s. 6d.

Alexander (T.).—ELEMENTARY APPLIED MECHANICS. Being the simpler and more practical Cases of Stress and Strain wrought out individually from first principles by means of Elementary Mathematics. By T. ALEXANDER, C.E., Professor of Civil Engineering in the Imperial College of Engineering, Tokei, Japan. Crown 8vo. Part I. 4s. 6d.

Alexander — Thomson. — ELEMENTARY APPLIED MECHANICS. By THOMAS ALEXANDER, C.E., Professor of Engineering in the Imperial College of Engineering, Tokei, Japan; and ARTHUR WATSON THOMSON, C.E., B.Sc., Professor of Engineering at the Royal College, Cirencester. Part II. TRANSVERSE STRESS; upwards of 150 Diagrams, and 200 Examples carefully worked out; new and complete method for finding, at every point of a beam, the amount of the greatest bending moment and shearing force during the transit of any set of loads fixed relatively to one another—*e.g.*, the wheels of a locomotive; continuous beams, &c., &c. Crown 8vo. 10s. 6d.

Awdry.—EASY LESSONS ON LIGHT. By Mrs. W. AWDRY. Illustrated. Extra fcap. 8vo. 2s. 6d.

Ball (R. S.).—EXPERIMENTAL MECHANICS. A Course of Lectures delivered at the Royal College of Science for Ireland. By R. S. BALL, M.A., Professor of Applied Mathematics and Mechanics in the Royal College of Science for Ireland. Cheaper Issue. Royal 8vo. 10s. 6d.

Chisholm. — THE SCIENCE OF WEIGHING AND MEASURING, AND THE STANDARDS OF MEASURE AND WEIGHT. By H.W. CHISHOLM, Warden of the Standards. With numerous Illustrations. Crown 8vo. 4s. 6d. (*Nature Series.*)

Clausius.—MECHANICAL THEORY OF HEAT. By R. CLAUSIUS. Translated by WALTER R. BROWNE, M.A., late Fellow of Trinity College, Cambridge. Crown 8vo. 10s. 6d.

Cotterill.—A TREATISE ON APPLIED MECHANICS. By JAMES COTTERILL, M.A., F.R.S., Professor of Applied Mechanics at the Royal Naval College, Greenwich. With Illustrations. 8vo. [*In the press.*

Cumming.—AN INTRODUCTION TO THE THEORY OF ELECTRICITY. By LINNÆUS CUMMING, M.A., one of the Masters of Rugby School. With Illustrations. Crown 8vo. 8s. 6d.

Daniell.—A TEXT-BOOK OF THE PRINCIPLES OF PHYSICS. By ALFRED DANIELL, M.A., Lecturer on Physics in the School of Medicine, Edinburgh. With Illustrations. Medium 8vo. 21s.

Day.—ELECTRIC LIGHT ARITHMETIC. By R. E. DAY, M.A., Evening Lecturer in Experimental Physics at King's College, London. Pott 8vo. 2s.

Everett.—UNITS AND PHYSICAL CONSTANTS. By J. D. EVERETT, F.R.S., Professor of Natural Philosophy, Queen's College, Belfast. Extra fcap. 8vo. 4s. 6d.

Gray.—ABSOLUTE MEASUREMENTS IN ELECTRICITY AND MAGNETISM. By ANDREW GRAY, M.A., F.R.S.E., Chief Assistant to the Professor of Natural History in the University of Glasgow. Pott 8vo. 3s. 6d.

Huxley.—INTRODUCTORY PRIMER OF SCIENCE. By T. H. HUXLEY, P.R.S., Professor of Natural History in the Royal School of Mines, &c. 18mo. 1s.

Kempe.—HOW TO DRAW A STRAIGHT LINE; a Lecture on Linkages. By A. B. KEMPE. With Illustrations. Crown 8vo. 1s. 6d. (*Nature Series.*)

Kennedy.—MECHANICS OF MACHINERY. By A. B. W. KENNEDY, M.Inst.C.E., Professor of Engineering and Mechanical Technology in University College, London. With Illustrations. Crown 8vo. [*In the press.*

Lang.—EXPERIMENTAL PHYSICS. By P. R. SCOTT LANG. M.A., Professor of Mathematics in the University of St. Andrews. Crown 8vo. [*In preparation.*

Martineau (Miss C. A.).—EASY LESSONS ON HEAT. By Miss C. A. MARTINEAU. Illustrated. Extra fcap. 8vo. 2s. 6d.

Mayer.—SOUND: a Series of Simple, Entertaining, and Inexpensive Experiments in the Phenomena of Sound, for the Use of Students of every age. By A. M. MAYER, Professor of Physics in the Stevens Institute of Technology, &c. With numerous Illustrations. Crown 8vo. 2s. 6d. (*Nature Series.*)

Mayer and Barnard.—LIGHT: a Series of Simple, Entertaining, and Inexpensive Experiments in the Phenomena of Light, for the Use of Students of every age. By A. M. MAYER and C. BARNARD. With numerous Illustrations. Crown 8vo. 2s. 6d. (*Nature Series.*)

Newton.—PRINCIPIA. Edited by Professor Sir W. THOMSON and Professor BLACKBURNE. 4to, cloth. 31s. 6d.

THE FIRST THREE SECTIONS OF NEWTON'S PRINCIPIA. With Notes and Illustrations. Also a Collection of Problems, principally intended as Examples of Newton's Methods. By PERCIVAL FROST, M.A. Third Edition. 8vo. 12s.

Parkinson.—A TREATISE ON OPTICS. By S. PARKINSON, D.D., F.R.S., Tutor and Prælector of St. John's College, Cambridge. New Edition, revised and enlarged. Crown 8vo. 10s. 6d.

Perry.—STEAM. AN ELEMENTARY TREATISE. By JOHN PERRY, C.E., Whitworth Scholar, Fellow of the Chemical Society, Lecturer in Physics at Clifton College. With numerous Woodcuts and Numerical Examples and Exercises. 18mo. 4s. 6d.

Ramsay.—EXPERIMENTAL PROOFS OF CHEMICAL THEORY FOR BEGINNERS. By WILLIAM RAMSAY, Ph.D., Professor of Chemistry in University College, Bristol. Pott 8vo. 2s. 6d.

Rayleigh.—THE THEORY OF SOUND. By LORD RAYLEIGH, M.A., F.R.S., formerly Fellow of Trinity College, Cambridge, 8vo. Vol. I. 12s. 6d. Vol. II. 12s. 6d.

[Vol. III. *in the press.*

Reuleaux.—THE KINEMATICS OF MACHINERY. Outlines of a Theory of Machines. By Professor F. REULEAUX. Translated and Edited by Professor A. B. W. KENNEDY, C.E. With 450 Illustrations. Medium 8vo. 21s.

Shann.—AN ELEMENTARY TREATISE ON HEAT, IN RELATION TO STEAM AND THE STEAM-ENGINE. By G. SHANN, M.A. With Illustrations. Crown 8vo. 4s. 6d.

Spottiswoode.—POLARISATION OF LIGHT. By the late W. SPOTTISWOODE, P.R.S. With many Illustrations. New Edition. Crown 8vo. 3s. 6d. (*Nature Series.*)

Stewart (Balfour).—Works by BALFOUR STEWART, F.R.S., Professor of Natural Philosophy in the Victoria University the Owens College, Manchester.

PRIMER OF PHYSICS. With numerous Illustrations. New Edition, with Questions. 18mo. 1s. (*Science Primers.*)

LESSONS IN ELEMENTARY PHYSICS. With numerous Illustrations and Chromolitho of the Spectra of the Sun, Stars, and Nebulæ. New Edition. Fcap. 8vo. 4s. 6d.

c

Stewart (Balfour).—Works by (*continued*)—
QUESTIONS ON BALFOUR STEWART'S ELEMENTARY LESSONS IN PHYSICS. By Prof. THOMAS H. CORE, Owens College, Manchester. Fcap. 8vo. 2s.

Stewart—Gee.—PRACTICAL PHYSICS, ELEMENTARY LESSONS IN. By Professor BALFOUR STEWART, F.R.S., and W. HALDANE GEE. Fcap. 8vo.

Part I. General Physics. [*Nearly ready.*
Part II. Optics, Heat, and Sound. [*In preparation.*
Part III. Electricity and Magnetism. [*In preparation.*

Stokes.—THE NATURE OF LIGHT. Burnett Lectures. By Prof. G. G. STOKES, Sec. R.S., etc. Crown 8vo. 2s. 6d.
ON LIGHT. Burnett Lectures. First Course. ON THE NATURE OF LIGHT. Delivered in Aberdeen in November 1883. By GEORGE GABRIEL STOKES, M.A., F.R.S., &c., Fellow of Pembroke College, and Lucasian Professor of Mathematics in the University of Cambridge. Crown 8vo. 2s. 6d.

Stone.—AN ELEMENTARY TREATISE ON SOUND. By W. H. STONE, M.B. With Illustrations. 18mo. 3s. 6d.

Tait.—HEAT. By P. G. TAIT, M.A., Sec. R.S.E., Formerly Fellow of St. Peter's College, Cambridge, Professor of Natural Philosophy in the University of Edinburgh. Crown 8vo. 6s.

Thompson.—ELEMENTARY LESSONS IN ELECTRICITY AND MAGNETISM. By SILVANUS P. THOMPSON. Professor of Experimental Physics in University College, Bristol. With Illustrations. Fcap. 8vo. 4s. 6d.

Thomson.—THE MOTION OF VORTEX RINGS, A TREATISE ON. An Essay to which the Adams Prize was adjudged in 1882 in the University of Cambridge. By J. J. THOMSON, Fellow and Assistant-Lecturer of Trinity College, Cambridge. With Diagrams. 8vo. 6s.

Todhunter.—NATURAL PHILOSOPHY FOR BEGINNERS. By I. TODHUNTER, M.A., F.R.S., D.Sc.
Part I. The Properties of Solid and Fluid Bodies. 18mo. 3s. 6d.
Part II. Sound, Light, and Heat. 18mo. 3s. 6d.

Wright (Lewis).— LIGHT; A COURSE OF EXPERIMENTAL OPTICS, CHIEFLY WITH THE LANTERN. By LEWIS WRIGHT. With nearly 200 Engravings and Coloured Plates. Crown 8vo. 7s. 6d.

ASTRONOMY.

Airy.—POPULAR ASTRONOMY. With Illustrations by Sir G. B. AIRY, K.C.B., formerly Astronomer-Royal. New Edition. 18mo. 4s. 6d.

Forbes.—TRANSIT OF VENUS. By G. FORBES, M.A., Professor of Natural Philosophy in the Andersonian University, Glasgow. Illustrated. Crown 8vo. 3s. 6d. (*Nature Series.*)

Godfray.—Works by HUGH GODFRAY, M.A., Mathematical Lecturer at Pembroke College, Cambridge.

A TREATISE ON ASTRONOMY, for the Use of Colleges and Schools. New Edition. 8vo. 12s. 6d.

AN ELEMENTARY TREATISE ON THE LUNAR THEORY, with a Brief Sketch of the Problem up to the time of Newton. Second Edition, revised. Crown 8vo. 5s. 6d.

Lockyer.—Works by J. NORMAN LOCKYER, F.R.S.

PRIMER OF ASTRONOMY. With numerous Illustrations. 18mo. 1s. (*Science Primers.*)

ELEMENTARY LESSONS IN ASTRONOMY. With Coloured Diagram of the Spectra of the Sun, Stars, and Nebulæ, and numerous Illustrations. New Edition. Fcap. 8vo. 5s. 6d.

QUESTIONS ON LOCKYER'S ELEMENTARY LESSONS IN ASTRONOMY. For the Use of Schools. By JOHN FORBES-ROBERTSON. 18mo, cloth limp. 1s. 6d.

THE SPECTROSCOPE AND ITS APPLICATIONS. With Coloured Plate and numerous Illustrations. New Edition. Crown 8vo. 3s. 6d.

Newcomb.—POPULAR ASTRONOMY. By S. NEWCOMB, LL.D., Professor U.S. Naval Observatory. With 112 Illustrations and 5 Maps of the Stars. Second Edition, revised. 8vo. 18s.

"It is unlike anything else of its kind, and will be of more use in circulating a knowledge of Astronomy than nine-tenths of the books which have appeared on the subject of late years."—SATURDAY REVIEW.

CHEMISTRY.

Fleischer.—A SYSTEM OF VOLUMETRIC ANALYSIS. Translated, with Notes and Additions, from the Second German Edition, by M. M. PATTISON MUIR, F.R.S.E. With Illustrations. Crown 8vo. 7s. 6d.

Jones.—Works by FRANCIS JONES, F.R.S.E., F.C.S., Chemical Master in the Grammar School, Manchester.
THE OWENS COLLEGE JUNIOR COURSE OF PRACTICAL CHEMISTRY. With Preface by Professor ROSCOE, and Illustrations. New Edition. 18mo. 2s. 6d.
QUESTIONS ON CHEMISTRY. A Series of Problems and Exercises in Inorganic and Organic Chemistry. Fcap. 8vo. 3s.

Landauer.—BLOWPIPE ANALYSIS. By J. LANDAUER. Authorised English Edition by J. TAYLOR and W. E. KAY, of Owens College, Manchester. Extra fcap. 8vo. 4s. 6d.

Lupton.—ELEMENTARY CHEMICAL ARITHMETIC. With 1,100 Problems. By SYDNEY LUPTON, M.A., Assistant-Master at Harrow. Extra fcap. 8vo. 5s.

Muir.—PRACTICAL CHEMISTRY FOR MEDICAL STUDENTS. Specially arranged for the first M.B. Course. By M. M. PATTISON MUIR, F.R.S.E. Fcap. 8vo. 1s. 6d.

Roscoe.—Works by H. E. ROSCOE, F.R.S. Professor of Chemistry in the Victoria University the Owens College, Manchester.
PRIMER OF CHEMISTRY. With numerous Illustrations. New Edition. With Questions. 18mo. 1s. (*Science Primers*).
LESSONS IN ELEMENTARY CHEMISTRY, INORGANIC AND ORGANIC. With numerous Illustrations and Chromolitho of the Solar Spectrum, and of the Alkalies and Alkaline Earths. New Edition. Fcap. 8vo. 4s. 6d.
A SERIES OF CHEMICAL PROBLEMS, prepared with Special Reference to the foregoing, by T. E. THORPE, Ph.D., Professor of Chemistry in the Yorkshire College of Science, Leeds, Adapted for the Preparation of Students for the Government, Science, and Society of Arts Examinations. With a Preface by Professor ROSCOE, F.R.S. New Edition, with Key. 18mo. 2s.

Roscoe and Schorlemmer.—INORGANIC AND ORGANIC CHEMISTRY. A Complete Treatise on Inorganic and Organic Chemistry. By Professor H. E. ROSCOE, F.R.S., and Professor C. SCHORLEMMER, F.R.S. With numerous Illustrations. Medium 8vo.
Vols. I. and II.—INORGANIC CHEMISTRY.
Vol. I.—The Non-Metallic Elements. 21s. Vol. II. Part I.—Metals. 18s. Vol. II. Part II.—Metals. 18s.
Vol. III.—ORGANIC CHEMISTRY. Two Parts.
THE CHEMISTRY OF THE HYDROCARBONS and their Derivatives, or ORGANIC CHEMISTRY. With numerous Illustrations. Medium 8vo. 21s. each.

SCIENCE.

Schorlemmer.—A MANUAL OF THE CHEMISTRY OF THE CARBON COMPOUNDS, OR ORGANIC CHEMISTRY. By C. SCHORLEMMER, F.R.S., Professor of Chemistry in the Victoria University the Owens College, Manchester. With Illustrations. 8vo. 14s.

Thorpe.—A SERIES OF CHEMICAL PROBLEMS, prepared with Special Reference to Professor Roscoe's Lessons in Elementary Chemistry, by T. E. THORPE, Ph.D., Professor of Chemistry in the Yorkshire College of Science, Leeds, adapted for the Preparation of Students for the Government, Science, and Society of Arts Examinations. With a Preface by Professor ROSCOE. New Edition, with Key. 18mo. 2s.

Thorpe and Rücker.—A TREATISE ON CHEMICAL PHYSICS. By Professor THORPE, F.R.S., and Professor RÜCKER, of the Yorkshire College of Science. Illustrated. 8vo. [*In preparation.*

Wright.—METALS AND THEIR CHIEF INDUSTRIAL APPLICATIONS. By C. ALDER WRIGHT, D.Sc., &c., Lecturer on Chemistry in St. Mary's Hospital Medical School. Extra fcap. 8vo. 3s. 6d.

BIOLOGY.

Allen.—ON THE COLOUR OF FLOWERS, as Illustrated in the British Flora. By GRANT ALLEN. With Illustrations. Crown 8vo. 3s. 6d. (*Nature Series.*)

Balfour.—A TREATISE ON COMPARATIVE EMBRYOLOGY. By F. M. BALFOUR, M.A., F.R.S., Fellow and Lecturer of Trinity College, Cambridge. With Illustrations. In 2 vols. 8vo. Vol. I. 18s. Vol. II. 21s.

Bettany.—FIRST LESSONS IN PRACTICAL BOTANY. By G. T. BETTANY, M.A., F.L.S., Lecturer in Botany at Guy's Hospital Medical School. 18mo. 1s.

Darwin (Charles).—MEMORIAL NOTICES OF CHARLES DARWIN, F.R.S., &c. By Professor HUXLEY, P.R.S., G. J. ROMANES, F.R.S., ARCHIBALD GEIKIE, F.R.S., and W. T. THISELTON DYER, F.R.S. Reprinted from *Nature*. With a Portrait, engraved by C. H. JEENS. Crown 8vo. 2s. 6d. (*Nature Series.*)

Dyer and Vines.—THE STRUCTURE OF PLANTS. By Professor THISELTON DYER, F.R.S., assisted by SYDNEY VINES, D.Sc., Fellow and Lecturer of Christ's College, Cambridge, and F. O. BOWER, M.A., Lecturer in the Normal School of Science. With numerous Illustrations. [*In preparation.*

Flower (W. H.)—AN INTRODUCTION TO THE OSTEOLOGY OF THE MAMMALIA. Being the substance of the Course of Lectures delivered at the Royal College of Surgeons of England in 1870. By Professor W. H. FLOWER, F.R.S., F.R.C.S. With numerous Illustrations. New Edition, enlarged. Crown 8vo. 10s. 6d.

Foster.—Works by MICHAEL FOSTER, M.D., F.R.S., Professor of Physiology in the University of Cambridge.
PRIMER OF PHYSIOLOGY. With numerous Illustrations. New Edition. 18mo. 1s.
A TEXT-BOOK OF PHYSIOLOGY. With Illustrations. Fourth Edition, revised. 8vo. 21s.

Foster and Balfour.—THE ELEMENTS OF EMBRYOLOGY. By MICHAEL FOSTER, M.A., M.D., LL.D., F.R.S., Professor of Physiology in the University of Cambridge, Fellow of Trinity College, Cambridge, and the late FRANCIS M. BALFOUR, M.A., LL.D., F.R.S., Fellow of Trinity College, Cambridge, and Professor of Animal Morphology in the University. Second Edition, revised. Edited by ADAM SEDGWICK, M.A., Fellow and Assistant Lecturer of Trinity College, Cambridge, and WALTER HEAPE, Demonstrator in the Morphological Laboratory of the University of Cambridge. With Illustrations. Crown 8vo. 10s. 6d.

Foster and Langley.—A COURSE OF ELEMENTARY PRACTICAL PHYSIOLOGY. By Prof. MICHAEL FOSTER, M.D., F.R.S., &c., and J. N. LANGLEY, M.A., F.R.S., Fellow of Trinity College, Cambridge. Fifth Edition. Crown 8vo. 7s. 6d.

Gamgee.—A TEXT-BOOK OF THE PHYSIOLOGICAL CHEMISTRY OF THE ANIMAL BODY. Including an Account of the Chemical Changes occurring in Disease. By A. GAMGEE, M.D., F.R.S., Professor of Physiology in the Victoria University the Owens College, Manchester. 2 Vols. 8vo. With Illustrations. Vol. I. 18s. [*Vol. II. in the press.*

Gegenbaur.—ELEMENTS OF COMPARATIVE ANATOMY. By Professor CARL GEGENBAUR. A Translation by F. JEFFREY BELL, B.A. Revised with Preface by Professor E. RAY LANKESTER, F.R.S. With numerous Illustrations. 8vo. 21s.

SCIENCE.

Gray.—STRUCTURAL BOTANY, OR ORGANOGRAPHY ON THE BASIS OF MORPHOLOGY. To which are added the principles of Taxonomy and Phytography, and a Glossary of Botanical Terms. By Professor ASA GRAY, LL.D. 8vo. 10s. 6d.

Hooker.—Works by Sir J. D. HOOKER, K.C.S.I., C.B., M.D., F.R.S., D.C.L.
PRIMER OF BOTANY. With numerous Illustrations. New Edition. 18mo. 1s. (*Science Primers*.)
THE STUDENT'S FLORA OF THE BRITISH ISLANDS. New Edition, revised. Globe 8vo. 10s. 6d.

Huxley.—Works by Professor HUXLEY, P.R.S.
INTRODUCTORY PRIMER OF SCIENCE. 18mo. 1s. (*Science Primers*.)
LESSONS IN ELEMENTARY PHYSIOLOGY. With numerous Illustrations. New Edition. Fcap. 8vo. 4s. 6d.
QUESTIONS ON HUXLEY'S PHYSIOLOGY FOR SCHOOLS. By T. ALCOCK, M.D. 18mo. 1s. 6d.
PRIMER OF ZOOLOGY. 18mo. (*Science Primers*.)
[*In preparation.*

Huxley and Martin.—A COURSE OF PRACTICAL INSTRUCTION IN ELEMENTARY BIOLOGY. By Professor HUXLEY, P.R.S., assisted by H. N. MARTIN, M.B., D.Sc. New Edition, revised. Crown 8vo. 6s.

Lankester.—Works by Professor E. RAY LANKESTER, F.R.S.
A TEXT BOOK OF ZOOLOGY. Crown 8vo. [*In preparation.*
DEGENERATION : A CHAPTER IN DARWINISM. Illustrated. Crown 8vo. 2s. 6d. (*Nature Series.*)

Lubbock.—Works by SIR JOHN LUBBOCK, M.P., F.R.S., D.C.L.
THE ORIGIN AND METAMORPHOSES OF INSECTS. With numerous Illustrations. New Edition. Crown 8vo. 3s. 6d. (*Nature Series.*)
ON BRITISH WILD FLOWERS CONSIDERED IN RELATION TO INSECTS. With numerous Illustrations. New Edition. Crown 8vo. 4s. 6d. (*Nature Series*).

M'Kendrick.—OUTLINES OF PHYSIOLOGY IN ITS RELATIONS TO MAN. By J. G. M'KENDRICK, M.D., F.R.S.E. With Illustrations. Crown 8vo. 12s. 6d.

Martin and Moale.—ON THE DISSECTION OF VERTEBRATE ANIMALS. By Professor H. N. MARTIN and W. A. MOALE. Crown 8vo. [*In preparation.*
(See also page 22.)

Miall.—STUDIES IN COMPARATIVE ANATOMY.
  No. I.—The Skull of the Crocodile: a Manual for Students. By L. C. MIALL, Professor of Biology in the Yorkshire College and Curator of the Leeds Museum. 8vo. 2s. 6d.
  No. II.—Anatomy of the Indian Elephant. By L. C. MIALL and F. GREENWOOD. With Illustrations. 8vo. 5s.

Mivart.—Works by ST. GEORGE MIVART, F.R.S. Lecturer in Comparative Anatomy at St. Mary's Hospital.
  LESSONS IN ELEMENTARY ANATOMY. With upwards of 400 Illustrations. Fcap. 8vo. 6s. 6d.
  THE COMMON FROG. With numerous Illustrations. Crown 8vo. 3s. 6d. (*Nature Series.*)

Müller.—THE FERTILISATION OF FLOWERS. By Professor HERMANN MÜLLER. Translated and Edited by D'ARCY W. THOMPSON, B.A., Scholar of Trinity College, Cambridge. With a Preface by CHARLES DARWIN, F.R.S. With numerous Illustrations. Medium 8vo. 21s.

Oliver.—Works by DANIEL OLIVER, F.R.S., &c., Professor of Botany in University College, London, &c.
  FIRST BOOK OF INDIAN BOTANY. With numerous Illustrations. Extra fcap. 8vo. 6s. 6d.
  LESSONS IN ELEMENTARY BOTANY. With nearly 200 Illustrations. New Edition. Fcap. 8vo. 4s. 6d.

Parker.—A COURSE OF INSTRUCTION IN ZOOTOMY (VERTEBRATA). By T. JEFFREY PARKER, B.Sc. London, Professor of Biology in the University of Otago, New Zealand. With Illustrations. Crown 8vo. 8s. 6d.

Parker and Bettany.—THE MORPHOLOGY OF THE SKULL. By Professor PARKER and G. T. BETTANY. Illustrated. Crown 8vo. 10s. 6d.

Romanes.—THE SCIENTIFIC EVIDENCES OF ORGANIC EVOLUTION. By G. J. ROMANES, M.A., LL.D., F.R.S., Zoological Secretary to the Linnean Society. Crown 8vo. 2s. 6d. (*Nature Series.*)

Smith.—Works by JOHN SMITH, A.L.S., &c.
  A DICTIONARY OF ECONOMIC PLANTS. Their History, Products, and Uses. 8vo. 14s.
  DOMESTIC BOTANY: An Exposition of the Structure and Classification of Plants, and their Uses for Food, Clothing, Medicine, and Manufacturing Purposes. With Illustrations. **New Issue.** Crown 8vo. 12s. 6d.

MEDICINE.

Brunton.—Works by T. LAUDER BRUNTON, M.D., Sc.D., F.R.C.P., F.R.S., Examiner in Materia Medica in the University of London, late Examiner in Materia Medica in the University of Edinburgh, and the Royal College of Physicians, London.

A TREATISE ON MATERIA MEDICA. 8vo. [*In the press.*

TABLES OF MATERIA MEDICA: A Companion to the Materia Medica Museum. With Illustrations. New Edition Enlarged. 8vo. 10s. 6d.

Hamilton.—A TEXT-BOOK OF PATHOLOGY. By D. J. HAMILTON, Professor of Pathological Anatomy (Sir Erasmus Wilson Chair), University of Aberdeen. 8vo. [*In preparation.*

Ziegler-Macalister.—TEXT-BOOK OF PATHOLOGICAL ANATOMY AND PATHOGENESIS. By Professor ERNST ZIEGLER of Tübingen. Translated and Edited for English Students by DONALD MACALISTER, M.A., M.B., B.Sc., M.R.C.P., Fellow and Medical Lecturer of St. John's College, Cambridge. With numerous Illustrations. Medium 8vo. Part I.—GENERAL PATHOLOGICAL ANATOMY. 12s. 6d.
Part II.—SPECIAL PATHOLOGICAL ANATOMY. Sections I.—VIII. 12s. 6d. [PART III. *in preparation.*

ANTHROPOLOGY.

Flower.—FASHION IN DEFORMITY, as Illustrated in the Customs of Barbarous and Civilised Races. By Professor FLOWER, F.R.S., F.R.C.S. With Illustrations. Crown 8vo. 2s. 6d. (*Nature Series*).

Tylor.—ANTHROPOLOGY. An Introduction to the Study of Man and Civilisation. By E. B. TYLOR, D.C.L., F.R.S. With numerous Illustrations. Crown 8vo. 7s. 6d.

PHYSICAL GEOGRAPHY & GEOLOGY.

Blanford.—THE RUDIMENTS OF PHYSICAL GEOGRAPHY FOR THE USE OF INDIAN SCHOOLS; with a Glossary of Technical Terms employed. By H. F. BLANFORD, F.R.S. New Edition, with Illustrations. Globe 8vo. 2s. 6d.

Geikie.—Works by ARCHIBALD GEIKIE, F.R.S., Director General of the Geological Surveys of the United Kingdom.
PRIMER OF PHYSICAL GEOGRAPHY. With numerous Illustrations. New Edition. With Questions. 18mo. 1s. (*Science Primers.*)
ELEMENTARY LESSONS IN PHYSICAL GEOGRAPHY. With numerous Illustrations. Fcap. 8vo. 4s. 6d.
QUESTIONS ON THE SAME. 1s. 6d.
PRIMER OF GEOLOGY. With numerous Illustrations. New Edition. 18mo. 1s. (*Science Primers.*)
ELEMENTARY LESSONS IN GEOLOGY. With Illustrations. Fcap. 8vo. [*In preparation.*
TEXT-BOOK OF GEOLOGY. With numerous Illustrations. 8vo. 28s.
OUTLINES OF FIELD GEOLOGY. With Illustrations. New Edition. Extra fcap. 8vo. 3s. 6d.

Huxley.—PHYSIOGRAPHY. An Introduction to the Study of Nature. By Professor HUXLEY, P.R.S. With numerous Illustrations, and Coloured Plates. New and Cheaper Edition. Crown 8vo. 6s.

AGRICULTURE.

Frankland.—AGRICULTURAL CHEMICAL ANALYSIS, A Handbook of. By PERCY FARADAY FRANKLAND, Ph.D., B.Sc., F.C.S., Associate of the Royal School of Mines, and Demonstrator of Practical and Agricultural Chemistry in the Normal School of Science and Royal School of Mines, South Kensington Museum. Founded upon *Leitfaden für die Agricultur-Chemische Analyse*, von Dr. F. KROCKER. Crown 8vo. 7s. 6d.

Tanner.—Works by HENRY TANNER, F.C.S., M.R.A.C., Examiner in the Principles of Agriculture under the Government Department of Science; Director of Education in the Institute of Agriculture, South Kensington, London; sometime Professor of Agricultural Science, University College, Aberystwith.
ELEMENTARY LESSONS IN THE SCIENCE OF AGRICULTURAL PRACTICE. Fcap. 8vo. 3s. 6d.
FIRST PRINCIPLES OF AGRICULTURE. 18mo. 1s.
THE PRINCIPLES OF AGRICULTURE. A Series of Reading Books for use in Elementary Schools. Prepared by HENRY TANNER, F.C.S., M.R.A.C. Extra fcap. 8vo.
  I. The Alphabet of the Principles of Agriculture. 6d.
  II. Further Steps in the Principles of Agriculture. 1s.
  III. Elementary School Readings on the Principles of Agriculture for the third stage. 1s.

POLITICAL ECONOMY.

Cossa.—GUIDE TO THE STUDY OF POLITICAL ECONOMY. By Dr. LUIGI COSSA, Professor in the University of Pavia. Translated from the Second Italian Edition. With a Preface by W. STANLEY JEVONS, F.R.S. Crown 8vo. 4s. 6d.

Fawcett (Mrs.).—Works by MILLICENT GARRETT FAWCETT:—
POLITICAL ECONOMY FOR BEGINNERS, WITH QUESTIONS. Fourth Edition. 18mo. 2s. 6d.
TALES IN POLITICAL ECONOMY. Crown 8vo. 3s.

Fawcett.—A MANUAL OF POLITICAL ECONOMY. By Right Hon. HENRY FAWCETT, M.P., F.R.S. Sixth Edition, revised, with a chapter on "State Socialism and the Nationalisation of the Land," and an Index. Crown 8vo. 12s.

Jevons.—PRIMER OF POLITICAL ECONOMY. By W. STANLEY JEVONS, LL.D., M.A., F.R.S. New Edition. 18mo. 1s. (Science Primers.)

Marshall.—THE ECONOMICS OF INDUSTRY. By A. MARSHALL, M.A., late Principal of University College, Bristol, and MARY P. MARSHALL, late Lecturer at Newnham Hall, Cambridge. Extra fcap. 8vo. 2s. 6d.

Sidgwick.—THE PRINCIPLES OF POLITICAL ECONOMY. By Professor HENRY SIDGWICK, M.A., Prælector in Moral and Political Philosophy in Trinity College, Cambridge, &c., Author of "The Methods of Ethics." 8vo. 16s.

Walker.—POLITICAL ECONOMY. By FRANCIS A. WALKER, M.A., Ph.D., Author of "The Wages Question," "Money," "Money in its Relation to Trade," &c. 8vo. 10s. 6d.

MENTAL & MORAL PHILOSOPHY.

Caird.—MORAL PHILOSOPHY, An Elementary Treatise on. By Prof. E. CAIRD, of Glasgow University. Fcap. 8vo.
[*In preparation.*

Calderwood.—HANDBOOK OF MORAL PHILOSOPHY. By the Rev. HENRY CALDERWOOD, LL.D., Professor of Moral Philosophy, University of Edinburgh. New Edition. Crown 8vo. 6s.

Clifford.—SEEING AND THINKING. By the late Professor W. K. CLIFFORD, F.R.S. With Diagrams. Crown 8vo. 3s. 6d. (*Nature Series.*)

Jevons.—Works by the late W. STANLEY JEVONS, LL.D., M.A., F.R.S.
> PRIMER OF LOGIC. New Edition. 18mo. 1s. (*Science Primers.*)
>
> ELEMENTARY LESSONS IN LOGIC; Deductive and Inductive, with copious Questions and Examples, and a Vocabulary of Logical Terms. New Edition. Fcap. 8vo. 3s. 6d.
>
> THE PRINCIPLES OF SCIENCE. A Treatise on Logic and Scientific Method. New and Revised Edition. Crown 8vo. 12s. 6d.
>
> STUDIES IN DEDUCTIVE LOGIC. Crown 8vo. 6s.

Keynes.—FORMAL LOGIC, Studies and Exercises in. Including a Generalisation of Logical Processes in their application to Complex Inferences. By JOHN NEVILLE KEYNES, M.A., late Fellow of Pembroke College, Cambridge. Crown 8vo. 10s. 6d.

Robertson.—ELEMENTARY LESSONS IN PSYCHOLOGY. By G. CROOM ROBERTSON, Professor of Mental Philosophy, &c., University College, London. [*In preparation.*

Sidgwick.—THE METHODS OF ETHICS. By Professor HENRY SIDGWICK, M.A., Prælector in Moral and Political Philosophy in Trinity College, Cambridge, &c. Second Edition. 8vo. 14s.

HISTORY AND GEOGRAPHY.

Arnold.—THE ROMAN SYSTEM OF PROVINCIAL ADMINISTRATION TO THE ACCESSION OF CONSTANTINE THE GREAT. By W. T. ARNOLD, B.A. Crown 8vo. 6s.
> "Ought to prove a valuable handbook to the student of Roman history."—GUARDIAN.

Beesly.—STORIES FROM THE HISTORY OF ROME. By Mrs. BEESLY. Fcap. 8vo. 2s. 6d.
> "The attempt appears to us in every way successful. The stories are interesting in themselves, and are told with perfect simplicity and good feeling." — DAILY NEWS.

Brook.—FRENCH HISTORY FOR ENGLISH CHILDREN. By SARAH BROOK. With Coloured Maps. Crown 8vo. 6s.

HISTORY AND GEOGRAPHY.

Clarke.—CLASS-BOOK OF GEOGRAPHY. By C. B. CLARKE, M.A., F.L.S., F.G.S., F.R.S. New Edition, with Eighteen Coloured Maps. Fcap. 8vo. 3s.

Freeman.—OLD-ENGLISH HISTORY. By EDWARD A. FREEMAN, D.C.L., LL.D., late Fellow of Trinity College, Oxford. With Five Coloured Maps. New Edition. Extra fcap. 8vo. 6s.

Fyffe.—A SCHOOL HISTORY OF GREECE. By C. A. FYFFE, M.A., Fellow of University College, Oxford. Crown 8vo. [*In preparation.*

Green.—Works by JOHN RICHARD GREEN, M.A., LL.D., late Honorary Fellow of Jesus College, Oxford.

SHORT HISTORY OF THE ENGLISH PEOPLE. With Coloured Maps, Genealogical Tables, and Chronological Annals. Crown 8vo. 8s. 6d. Ninety-ninth Thousand.

"Stands alone as the one general history of the country, for the sake of which all others, if young and old are wise, will be speedily and surely set aside."—ACADEMY.

ANALYSIS OF ENGLISH HISTORY, based on Green's "Short History of the English People." By C. W. A. TAIT, M.A., Assistant-Master, Clifton College. Crown 8vo. 3s. 6d.

READINGS FROM ENGLISH HISTORY. Selected and Edited by JOHN RICHARD GREEN. Three Parts. Globe 8vo. 1s. 6d. each. I. Hengist to Cressy. II. Cressy to Cromwell. III. Cromwell to Balaklava.

A SHORT GEOGRAPHY OF THE BRITISH ISLANDS. By JOHN RICHARD GREEN and ALICE STOPFORD GREEN. With Maps. Fcap. 8vo. 3s. 6d.

Grove.—A PRIMER OF GEOGRAPHY. By Sir GEORGE GROVE, D.C.L., F.R.G.S. With Illustrations. 18mo. 1s. (*Science Primers.*)

Guest.—LECTURES ON THE HISTORY OF ENGLAND. By M. J. GUEST. With Maps. Crown 8vo. 6s.

"It is not too much to assert that this is one of the very best class books of English History for young students ever published."—SCOTSMAN.

Historical Course for Schools—Edited by EDWARD A. FREEMAN, D.C.L., late Fellow of Trinity College, Oxford.

I.—GENERAL SKETCH OF EUROPEAN HISTORY. By EDWARD A. FREEMAN, D.C.L. New Edition, revised and enlarged, with Chronological Table, Maps, and Index. 18mo. 3s. 6d.

Historical Course for Schools. *Continued—*

II.—HISTORY OF ENGLAND. By EDITH THOMPSON. New Edition, revised and enlarged, with Coloured Maps. 18mo. 2s. 6d.

III.—HISTORY OF SCOTLAND. By MARGARET MACARTHUR. New Edition. 18mo. 2s.

IV.—HISTORY OF ITALY. By the Rev. W. HUNT, M.A. New Edition, with Coloured Maps. 18mo. 3s. 6d.

V.—HISTORY OF GERMANY. By J. SIME, M.A. 18mo. 3s.

VI.—HISTORY OF AMERICA. By JOHN A. DOYLE. With Maps. 18mo. 4s. 6d.

VII.—EUROPEAN COLONIES. By E. J. PAYNE, M.A. With Maps. 18mo. 4s. 6d.

VIII.—FRANCE. By CHARLOTTE M. YONGE. With Maps. 18mo. 3s. 6d.

GREECE. By EDWARD A. FREEMAN, D.C.L. [*In preparation.*

ROME. By EDWARD A. FREEMAN, D.C.L. [*In preparation.*

History Primers—Edited by JOHN RICHARD GREEN, M.A., LL.D., Author of "A Short History of the English People."

ROME. By the Rev. M. CREIGHTON, M.A., late Fellow and Tutor of Merton College, Oxford. With Eleven Maps. 18mo. 1s.

"The author has been curiously successful in telling in an intelligent way the story of Rome from first to last."—SCHOOL BOARD CHRONICLE.

GREECE. By C. A. FYFFE, M.A., Fellow and late Tutor of University College, Oxford. With Five Maps. 18mo. 1s.

"We give our unqualified praise to this little manual."—SCHOOLMASTER.

EUROPEAN HISTORY. By E. A. FREEMAN, D.C.L., LL.D. With Maps. 18mo. 1s.

"The work is always clear, and forms a luminous key to European history." —SCHOOL BOARD CHRONICLE.

GREEK ANTIQUITIES. By the Rev. J. P. MAHAFFY, M.A. Illustrated. 18mo. 1s.

"All that is necessary for the scholar to know is told so compactly yet so fully, and in a style so interesting, that it is impossible for even the dullest boy to look on this little work in the same light as he regards his other school books."—SCHOOLMASTER.

CLASSICAL GEOGRAPHY. By H. F. TOZER, M.A. 18mo. 1s.

"Another valuable aid to the study of the ancient world. . . . It contains an enormous quantity of information packed into a small space, and at the same time communicated in a very readable shape."—JOHN BULL.

History Primers *Continued*—

GEOGRAPHY. By Sir GEORGE GROVE, D.C.L. With Maps. 18mo. 1s.

"A model of what such a work should be. . . . We know of no short treatise better suited to infuse life and spirit into the dull lists of proper names of which our ordinary class-books so often almost exclusively consist."—TIMES.

ROMAN ANTIQUITIES. By Professor WILKINS. Illustrated. 18mo. 1s.

"A little book that throws a blaze of light on Roman history, and is, moreover intensely interesting."—SCHOOL BOARD CHRONICLE.

FRANCE. By CHARLOTTE M. YONGE. 18mo. 1s.

"May be considered a wonderfully successful piece of work. . . . Its general merit as a vigorous and clear sketch, giving in a small space a vivid idea of the history of France, remains undeniable."—SATURDAY REVIEW.

Hole.—A GENEALOGICAL STEMMA OF THE KINGS OF ENGLAND AND FRANCE. By the Rev. C. HOLE. On Sheet. 1s.

Kiepert—A MANUAL OF ANCIENT GEOGRAPHY. From the German of Dr. H. KIEPERT. Crown 8vo. 5s.

Lethbridge.—A SHORT MANUAL OF THE HISTORY OF INDIA. With an Account of INDIA AS IT IS. The Soil, Climate, and Productions; the People, their Races, Religions, Public Works, and Industries; the Civil Services, and System of Administration. By ROPER LETHBRIDGE, M.A., C.I.E., late Scholar of Exeter College, Oxford, formerly Principal of Kishnaghur College, Bengal, Fellow and sometime Examiner of the Calcutta University. With Maps. Crown 8vo. 5s.

Michelet.—A SUMMARY OF MODERN HISTORY. Translated from the French of M. MICHELET, and continued to the Present Time, by M. C. M. SIMPSON. Globe 8vo. 4s. 6d.

Otté.—SCANDINAVIAN HISTORY. By E. C. OTTÉ. With Maps. Globe 8vo. 6s.

Ramsay.—A SCHOOL HISTORY OF ROME. By G. G. RAMSAY, M.A., Professor of Humanity in the University of Glasgow. With Maps. Crown 8vo. [*In preparation.*]

Tait.—ANALYSIS OF ENGLISH HISTORY, based on Green's "Short History of the English People." By C. W. A. TAIT, M.A., Assistant-Master, Clifton College. Crown 8vo. 3s. 6d.

Wheeler.—A SHORT HISTORY OF INDIA AND OF THE FRONTIER STATES OF AFGHANISTAN, NEPAUL, AND BURMA. By J. TALBOYS WHEELER. With Maps. Crown 8vo. 12s.

"It is the best book of the kind we have ever seen, and we recommend it to a place in every school library."—EDUCATIONAL TIMES.

Yonge (Charlotte M.).—A PARALLEL HISTORY OF FRANCE AND ENGLAND; consisting of Outlines and Dates. By CHARLOTTE M. YONGE, Author of "The Heir of Redclyffe," &c., &c. Oblong 4to. 3s. 6d.

CAMEOS FROM ENGLISH HISTORY.—FROM ROLLO TO EDWARD II. By the Author of "The Heir of Redclyffe." Extra fcap. 8vo. New Edition. 5s.

A SECOND SERIES OF CAMEOS FROM ENGLISH HISTORY. — THE WARS IN FRANCE. New Edition. Extra fcap. 8vo. 5s.

A THIRD SERIES OF CAMEOS FROM ENGLISH HISTORY. —THE WARS OF THE ROSES. New Edition. Extra fcap. 8vo. 5s.

CAMEOS FROM ENGLISH HISTORY—A FOURTH SERIES. REFORMATION TIMES. Extra fcap. 8vo. 5s.

CAMEOS FROM ENGLISH HISTORY.—A FIFTH SERIES. ENGLAND AND SPAIN. Extra fcap. 8vo. 5s.

EUROPEAN HISTORY. Narrated in a Series of Historical Selections from the Best Authorities. Edited and arranged by E. M. SEWELL and C. M. YONGE. First Series, 1003—1154. New Edition. Crown 8vo. 6s. Second Series, 1088—1228. New Edition. Crown 8vo. 6s.

www.ingramcontent.com/pod-product-compliance
Lightning Source LLC
Chambersburg PA
CBHW020831230426
43666CB00007B/1178